Introduction to the Philosophy of Nature

American University Studies

Series V
Philosophy

Vol. 95

PETER LANG
New York · San Francisco · Bern
Frankfurt am Main · Paris · London

Florence M. Hetzler

Introduction to the Philosophy of Nature

PETER LANG
New York · San Francisco · Bern
Frankfurt am Main · Paris · London

Library of Congress Cataloging-in-Publication Data

Hetzler, Florence M.
Introduction to the philosophy of nature / Florence M. Hetzler.
p. cm. – (American university studies. Series V, Philosophy ; vol. 95)
"The commentary of St. Thomas Aquinas on book one of the Physics of Aristotle"– P.
Includes bibliographical references.
1. Aristotle. Physics. Book 1. 2. Thomas, Aquinas, Saint, 1225?-1274. In octo libros Physicorum expositio. English. 1990. 3. Philosophy of nature – Early works to 1800. 4. Physics – Philosophy – Early works to 1800. 5. Science, Ancient. I. Thomas, Aquinas, Saint, 1225?-1274. In octo libros Physicorum expositio. English. 1990. II. Title. III. Series.
Q151.H47 1990 113 – dc20 90-21247
ISBN 0-8204-1278-3 CIP
ISSN 0739-6392

Printed in the United States of America.

To my parents

TABLE OF CONTENTS

THE COMMENTARY OF ST. THOMAS AQUINAS ON BOOK ONE OF THE PHYSICS OF ARISTOTLE

PREFACE

It would seem that the ideal manner of determining what a philosopher has said would be to study the works of that philosopher rather than what someone else has said about him in intellectual "hand-me-downs." For this reason, the intention in this work is to return to the original text of St. Thomas. The first book of St. Thomas' Commentary on the *Physics* of Aristotle is presented as a self-contained unit, as far as this is possible practically. In order that this study may be an integrated work the following are included:

1. The text of Aristotle. This is the Wicksteed-Cornford translation of the Greek into English. This text has been renumbered to correspond to the various citations to which St. Thomas referred throughout his work. It seems unwise to intersperse the translation of St. Thomas with incomplete quotations that refer to the Aristotelian text, unwise not only because these words are not placed in a significant sentence in the text of St. Thomas but also because these "half-quotes" are distracting to the reader.

2. The commentary of St. Thomas. This is my translation of the text as taken from *In Octo Libros Physicorum Aristotelis Expositio* prepared by T. M. Maggiolo and published by Marietti at Rome in 1954. I have tried to adhere as closely as possible to the text, while maintaining at the same time a reading fluidity which does not sacrifice philosophical content for linguistic style in this difficult work of St. Thomas.

3. Footnotes. These are of two kinds. They may be references or cross-references to a given citation or meaning, or they may include the pertinent quotes

of the Pre-Socratic philosophers. These latter have been taken from Kathleen Freeman's *Ancilla to the Pre-Socratic Philosophers* published in 1956 by Basil Blackwell in Oxford. Incorporation of these quotations seems a worthy addition for a thorough understanding of this work, for they concern the early natural philosophers who considered many of the same problems that Aristotle either solved or refuted.

5. The Appendix. This includes explanations of various terms or passages in the text that may be vague to one who is not familiar with them or which may be difficult because they are seemingly ambiguous, paradoxical or contradictory. These explanations provide either a clarification or exemplification made elsewhere by St. Thomas or others, or give my own explanation of the meaning of St. Thomas. At times the references or cross-references that are quoted may appear rather lengthy. The completeness of these quotes should, however, prove advantageous to the student who does not read Latin or who does not have ready access to these citations. The Appendix follows the completion of Book I in order that the continuity of the text of St. Thomas may not be interrupted.

INTRODUCTION

In bringing forth a new translation and an integrated text of St. Thomas' *Commentary on the First Book of Aristotle's Physics*, Professor Florence Hetzler has rendered a service of signal importance to the philosophic community. This is true for several reasons. The first, and of course the most obvious, is that it fills a long-felt need, since no completely annotated and cross-referenced translation of this important work exists. In addition, it is important to note, as Dr. Hetzler has carefully pointed out in her Preface, that the First Book of St. Thomas' Commentary may be studied with great profit as an independent, or at least quasi-independent, self-contained unit on the methodology of natural philosophy. This, to be sure, is not to be understood as an attempt to wrench it from its natural setting within the whole context of the Commentary, of which it is an integral part. Rather, the value of both Hetzler's translation of Thomas and her commentary on him is that of opening for the first time to students and scholars who do not read Latin the inexhaustible riches of the thought of Aquinas.

This work comes at a most opportune time, since, as is generally accepted among those philosophers who are involved with the Philosophy of Science, nearly all of the theories that have been generated thus far seem to have reached a kind of end point. The reason for this unfortunate impasse—or at least so students of Thomas' *Commentary on the Physics* would argue—is that Metaphysics has been too much separated from the Philosophy of Nature. This divorce of Metaphysics from Philosophy of Nature has had two consequences, both unfortunate. On the one hand, as the positive sciences, e.g. Mathematical Physics, have become more and more mathematized, and as eidetic meaning has been more and more extruded from them, they have increasingly tended to be severed from the world of nature which they are attempting to explain. They have become so completely empty that they have produced the results described so effectively by the late Charles de Koninck in *The Hollow Universe*. We are left with an empty latticework in which even the lattices are getting thinner and thinner. This insatiable drive towards a total mathematization of science that is a total emptying and a total schematization of science was noted by no less a powerful and penetrating philosophical mind in the twentieth century than Martin Heidegger, especially in *Die Frage nach dem Ding*.

While Heidegger's well-known critique of this mathematizing (that is to say schematizing and emptying drive of modern science) is certainly well taken, still Heidegger's thought, which in the estimation of even his sternest critics represents on of the most powerful attempts in the history of modern philosophy to reconstruct a viable ontology, in the final analysis falls victim to the opposite tendency. He quite correctly pointed out not only the limitations of the mathematizing spirit of modern science, and even the threat that it poses to the survival of man *as man*, but his ontology so far separated itself from any Philosophy of Nature, it has ended, at least so his critics would argue, by being so ethereal that it is more or less totally removed from the real world of day to day experience in which we must perforce live and move and have our being. And so from the lofty heights of Todtnauberg the late Heidegger might flash forth his Jovian jeremiads against modern science, only to end in an impenetrable Germanic mystic prose-poetry. And this is the second of the unfortunate consequences mentioned above that a divorce of Metaphysics from the Philosophy of Nature has.

One of the great services, therefore, that Professor Hetzler has rendered to the scholarly community by her translation of Thomas's Commentary on this First book of the *Physics* is to make available a treatise which proposes an alternative theory which is able to avoid the Scylla of science's all consuming passion for total mathematization on the one hand, and the Charybdis of the penchant of contemporary metaphysicians to go their own way, disdainful of the plebian labors of science on the other. As Dr. Hetzler points out, Thomas was able to do this because in this Commentary on the First Book he sets forth his theory of the method which he will employ in developing a Philosophy of Nature. The Philosophy of Nature in the overall philosophy of Aquinas will have a pivotal role, and what is important in terms of the contemporary problem of formulating an adequate philosophic base from which to work out theories of scientific methodologies, his theory of the philosophy of Nature will stand at the crossroads between science, as it is understood today, and metaphysics. As Hetzler notes in her Commentary, with his notion of the *Scientiae Mediae* Aquinas provides a theoretical structure which is of the greatest utility to contemporary scientific methodology and theory, and this for two reasons. The first of these is that the *Scientiae Mediae*, Optics for example for the medieval Philosopher, mathematical Physics for the contemporary, partake of both realms, the mathematical and the physical. It can do this because the *objectum formale quo*, the intelligible light by which it views that which it investigates, is mathematical, while the *objectum formale quod*, that which it investigates is physical. In a word, it participates in both domains, using tools which are essentially mathematical to investigate physical phenomena. This theory has the considerable advantage of preventing the contemporary *Scientiae Mediae* such as mathematical Physics from becoming so far removed from the world of nature as it is open to common experience that they finish by being merely the investigation of the progressively dissolving lattices of the universe, totally removed from the real world as we know it in our day to day unspecialized experience. Yet because in this theory these sciences are formally mathematical it is able to incorporate the tremendous heuristic power of these new mathematical tools. We have, so to say, the best of both possible worlds.

Further, as we noted above, for Aquinas the role of the Philosophy of Nature is to stand at the crossroads of what we would now call the positive or physical sciences and First Philosophy or Metaphysics. This also has the very distinct advantage of preventing metaphysics from losing contact with reality and building beautiful, ethereal abstract, Hegelian, but, be it carefully noted, unreal, metaphysical castles in the sky. This metaphysical temptation towards excessive and unreal abstractness, whether it be as it is expressed in the thought of a Plato, a Descartes, a Hegel, or whoever, it is successfully resisted because, regardless of how sublime the height of its metaphysical speculations, it remains rooted in the firm ground of Philosophy of Nature.

From these brief remarks we may see the tremendous importance of the work which Dr. Hetzler has done. She has made available for the first time a work of St. Thomas which is able to furnish a key to the solution of the several dilemmas alluded to above, both in contemporary science and also in metaphysics. Hetzler has done this by affording us a translation of Thomas' Commentary on Aristotle which is admirable in its clarity and also in its fidelity to the text of St. Thomas. She has succeeded in rendering not only the literal sense of Thomas, but above all in allowing his spirit to shine through her text with laudable faithfulness. These several virtues she has combined with a readable prose style. In addition to this she has given over more than half of the work to a Commentary on various points in the text of St. Thomas which might be otherwise obscure to those uninitiated in

Thomistic Philosophy. This will, of course, make the appeal of her work all the wider, since it can also be a most useful first introduction to Thomas' Philosophy of Nature to anyone interested in scientific methodology, even though he might be quite unacquainted with the general contours of Thomistic thought. And finally, though by no means least, she has provided a very extensive cross-referencing with other related texts of Thomas which throw light on otherwise obscure points, so that what we have is the very felicitous situation in which the Master himself is made to clarify points with which we might experience some difficulty. To all of this she has added an excellent bibliography.

Prof. Thomas A. Fay, Chairman
Department of Philosophy
St. John's University
Jamaica, New York

SUMMARY OF BOOK I

Assuming that there are three objectives accomplished in this first book of the *Commentary*, it is most interesting to follow briefly and simply the logical manner in which St. Thomas proceeds to the realization of these three goals.

The first is the establishment of the method of procedure for all the works of natural philosophy, namely, that of advancing from the confused to the distinct, from the general to the specific. And St. Thomas proceeds from the universal knowledge in the *In Physicorum* to the less universal knowledge of the *In De Coelo*, i.e., from the study of movement in general to the study of the movement of the celestial spheres.

Secondly, in his investigation of the nature and number of the principles of natural beings, following Aristotle Thomas first studies the opinions already set forth and then proceeds to tell why they are inadequate. He rejects Parmenides and Melissus who posited *one immobile principle* and he declares the belief of these men to be invalid because it rejects the testimony of the senses and also involves certain logical fallacies. It is contrary to the very essence of beings of nature, which are *mobile* by reason of their materiality and, therefore, must admit of change or multiplicity. Even the Platonists are brought in and rejected, since, according to St. Thomas, their theory that non-being is being was in line with the theories of Parmenides and Melissus. In his answer to them, St. Thomas underlines the fallacy involved in taking unity in a univocal rather than in an analogical sense. The theories of Anaximines, who posited air as a principle, Thales, who postulated water and Heraclitus fire are discarded next. Besides those who held *one mobile principle*, those philosophers were also proved wrong who believed in a *finite plurality of principles*, be it two, three or four, as in the case of Empedocles. Having removed from the carpet of truth either *one immobile* or one *mobile principle*, as well as the given *finite number* of principles, there remains the elimination of the theory of an *infinite* number of principles. Democritus' theory of infinite indivisibles or atoms and the belief of Anaxagoras in an infinite number of minimal parts as the elements from which all other beings come to be are also rejected.

Lessons ten through fifteen are dedicated to the more proximate resolution of the problem at hand. The tenth lesson considers how all the ancient philosophers held that there was contrariety in first principles: even those who held one immobile being posited diversity according to the senses; those who held one material mobile principle posited contrariety according to the rarefied and the dense; and those who held more than one, whether finite or infinite, invoked congregation and segregation. After making clear how and why the ancient philosophers held contrary principles and why, if principles are contraries, they cannot be either one only, or infinite, Thomas explains that there must be a subject as a persisting basis for the being and becoming of natural entities. This substrate, which passes from not having "such a form" to having this form, is *matter*. The *form* which it attains is the second principle and the lack of the form before it was informed by it is privation, the third principle, which, since it does not enter intrinsically into the being that comes to be, is an incidental principle in contradistinction to the other two principles, which are substantial.

Matter is not only capacity for the form which it is to receive but somehow tends to it in the sense of being-for-form. It differs from privation because privation is non-being, although it is non-being considered in a definite generic frame of reference. Matter is not non-being but it is potency for form, for fullness of being, that is, for its perfection in the order of being. Matter, form and privation may be one in subject but not in meaning. Thus, with the architectonic establishment of matter, form and privation as the true principles of natural being, the third goal of this book has been accomplished.

It seems that the *Physics* of Aristotle and Thomas' *Commentary* upon it are more pertinent today than they were even fifty years ago: there is a need to understand the finite and the infinite. These works have always given in a quintessential manner the meaning of finite being and have been an excellent prelude to the understanding of the infinite being or being without matter of Aristotle's *Metaphysics*. If one consistently rethinks the orchestration of the whole Aristotelian and Thomistic traditions, the place of the Physics becomes increasingly important. Mobile being, finite being, and fragility, are notions of very great concerns for us today. Man is finite; he makes war when he could make peace; he hates when he could love. Is it his finitude that can be blamed for this? I think not. There is a tending towards or a tending beyond in the being of finite man that has not been sufficiently examined in many philosophies of the human person. This is not only because there is something in man that is beyond the finite.

The greatness of the finite can be studied in the *Physics*. There is sheer dynamism in the notion of privation in the study of finite being; we are defined in large part by what we are not. Privation, though considered an accidental principle in the present work in the coming to be of something, is an indispensable and exciting principle of change. It is because of what we are as well as what we are not that we change. Without the what we are not there would be no what we are or what we could or might be. Nor would there be a what we were. Heraclitus saw this well and it is his theory perhaps more than that of the other Pre-Socratics, and even Plato and Aristotle, that makes this such a dynamic theory and such a dynamic and fecund possibility in the physical as well as the intentional or logical orders of being. It is privation that makes creativity possible in large part and it is privation that has not been given the status that it deserves. Privation's infinity is the basis for part of the great mystery and potentiality for creativity of finite being. In the *Physics* we see part of this as the author proceeds to the nature of mobile being, finite being, and proceeds down to the finitude of planets, plants and parts of animals in such works as the *De Caelo*, *De Plantis* and *De Partibus Animalium*.

The understanding of finite being that can be obtained in the *Physics* is important not only for the understanding of Metaphysics and natural theology and the philosophy of God but also of the finite being of technology, and fine art, too, that are extensions of man. Prof. Gerald McCool, a noted Thomist scholar, has said that the Thomistic tradition is one of being rethought and re-reflected upon. For him it is this reflection that constitutes its unity. Perhaps the place of the *Physics* in the entire Thomistic tradition should be rethought not only to see the place of the *Physics* in the tradition but the changes and enrichment that a deeper consideration of the *Physics* might effect within and upon it. The more we become involved with matter, the more we are involved with spirit. The more we study matter and its privations, the more we see its possibilities—and ours.

I present this work as one who has studied the *Physics* for decades. I was immersed in its study even in college. Now I am convinced even more of its immense relevance. Dare I say that it becomes increasingly more relevant to our situation? It is an aid not only to the understanding of who and what we are, but also of where we are.

Florence M. Hetzler
Fordham University
New York, New York

SUMMARY GUIDE TO SYMBOLS IN TEXT

1) Numerical symbols at beginning of each section: refer to numeration in Marietti edition.

2) Numerical symbols in parentheses: refer to text of Aristotle that precedes each lesson.

3) Raised alphabetic symbols: refer to footnotes at bottom of page.

4) Raised numerical symbols: refer to Appendix notes following Book One.

TEXT OF ARISTOTLE[a] **LESSON ONE**[b]

1. In all sciences that are concerned with principles or causes or elements, it is acquaintance with these that constitutes knowledge or understanding. For we conceive ourselves to know about a thing when we are acquainted with its ultimate causes and first principles, and have got down to its elements. Obviously, then, in the study of Nature too, our first object must be to establish principles.

2. Now the path of investigation must lie from what is more immediately cognizable and clear to us, to what is clearer and more intimately cognizable in its own nature; for it is not the same thing to be directly accessible to our cognition and to be intrinsically intelligible. Hence, in advancing to that which is intrinsically more luminous and by its nature accessible to deeper knowledge, we must needs start from what is more immediately within our cognition, though in its own nature less fully accessible to understanding.

2b. Now the things most obvious and immediately cognizable by us are concrete and particular, rather than abstract and general; whereas elements and principles are only accessible to us afterwards, as derived from the concrete data when we have analyzed them. So we must advance from the concrete whole to the several constituents which it embraces;

3. for it is the concrete whole that is the more readily cognizable by the senses. And by calling the concrete a 'whole' I mean that it embraces in a single complex a diversity of constituent elements, factors, or properties.

4. The relation of names to definitions will throw some light on this point; for the name gives an unanalysed indication of the thing ('circle,' for instance) but

---

[a]This translation is that of Philip H. Wicksteed and Francis M. Cornford in *Aristotle: The Physics* (New York: G. P. Putnam's Sons, 1929).

[b]Includes 11. 184 a 10—184 b 14, Wicksteed and Cornford, *op. cit.*, pp. 11–13.

the definition analyses out some characteristic property or properties. A variant of the same thing may be noted in children, who begin by calling every man 'father' and every woman 'mother,' till they learn to sever out the special relation to which the terms properly apply.

The matter and subject of natural philosophy and of this treatise. We should proceed from principles that are more universal and more known to us.

1. Because the *Physics*, which we wish to set forth here, is the first of the treatises of natural philosophy,[a] it is fitting to point out at the beginning of it *what the matter and what the subject of natural philosophy are*.[b,1]

We should know that since all science is in the intellect,[2] whatever becomes intelligible in act does so by being somehow abstracted from matter,[c,3] for things

---

[a]To avoid ambiguity we shall use the term "natural philosophy" or "philosophy of nature" instead of "natural science" which today commonly refers to experimental science only.

[b]Cf. John of St. Thomas, *Cursus Philosophicus*, ed. Reiser, Vol. II, qu. 1, art. 1, p. 8; *Cursus Theologicus*, ed. Solesmes, Vol. I, Ia, disp. 2, art. 11, nn. 1–2, p. 402; Cajetan, *In Summam Theologicam*, contained in *Opera Omnia Sancti Thomae*, ed. Leonina, Ia, qu. 1, art. 7, nn. 8–12.

[c]Cf. St. Thomas Aquinas, *Quaestiones Disputatae De Veritate*, ed. Marietti, qu. 2, art. 5, corp.; John of St. Thomas, *Curs. Theol.*, Vol. III, qu. 10, art. 1, pp. 295–304 and pp. 322–339; Normand Marcotte, "The Knowability of Matter Secundum Se," *Laval théologique et philosophique*, Vol. I, No. 1 (1945), pp. 103–118; Thomas Hébert, A.A., "Notre connaissance intellectuelle du singulier matériel," *ibid.*, Vol. V, No. 1 (1949), pp. 33–65; G. Klubertanz, S.J., "St. Thomas and the Knowledge of the Material Singular," *The New Scholasticism*, Vol. XXVI (1952), pp. 35–66; R. Allers, "The Intellectual Cognition of Particulars," *Thomist*, Vol. III (1941), pp. 95–163; Maurice Holloway, S.J., "Abstraction in Cognition," *The Modern Schoolman*, Vol. XXIII, No. 3 (March, 1946), pp. 120–130.

pertain to different sciences[4] to the extent that they are diversely referred to matter. Furthermore, since all science requires demonstration,[5] and since the definition is the middle term of a demonstration, sciences must differ in kind on the basis of the different modes of definition.[a,6]

2. We should also know that there are things which depend upon matter for their very being and which cannot be defined without matter. But there are others which, although they cannot be except in sensible matter, nevertheless, do not have sensible matter in their definition.[b] And these beings differ from one another just as do the curved and the snub. For the snub is in sensible matter and must have sensible matter in its definition, since a snub nose is a curved nose. And the same is true of all natural beings, e.g., man, or a stone; but the curved, although it cannot exist unless it be in sensible matter, does not, however, have sensible matter in its definition. And this holds true for all mathematical beings, e.g., numbers, magnitudes and figures. But there are still other things that do not depend upon

---

[a]On the specification of the sciences, cf.: St. Thomas, *In I Posteriorum Analyticorum*, less. 41, nn. 10–11; *In Librum Boethii De Trinitate*, qu. 5; John of St. Thomas, *Curs. Phil.*, Vol. I, Log. 2, qu. 27, art. 1, pp. 818–824; Bernard Mullahy, C.S.C., *Subalternation and Mathematical Physics*, unpublished thesis (Quebec: Laval University, 1947); Henri Pichette, "Quelques principes fondamentaux de la doctrine du spéculatif et du pratique," *Laval théol. et phil.*, Vol. I, n. 1 (1945), pp. 52–70; Jacques Maritain, *The Degrees of Knowledge* (New York: Scribner's, 1938); *Philosophy of Nature* (New York: Philosophical Library, 1951), pp. 73–157.

[b]On intelligible matter, cf.: St. Thomas, *In VII Metaphysicam Aristotelis*, less. 10, nn. 1494–1496; *ibid.*, less. 11, entire less.; *in VIII Metaph.*, less. 5, nn. 1760–1762; *In III De Anima*, less 8, nn. 707–715; *Summa Theologiae*, Ia, qu. 85, art. 1, ans. 2; *De Trin.*, qu. 5, art. 3, corp. and ans. 3, 4; *ibid.*, qu. 6, art. 2, corp. and ans. 2–5; *In II Post. Anal.*, less. 9, n. 5; *De Ver.*, qu. 2, art. 6, ans. 1.

matter either for their being or for their definition,[a] either because they are never in matter, e.g., God, and other separated substances, or because they are not universally in matter, e.g., substance, potency, act, and being itself.[b]

3. Metaphysics, therefore, is concerned with beings of this kind, while mathematics concerns those that depend upon sensible matter for their being but not for their definition. But that part of natural philosophy, which is called physics, concerns those things that depend upon matter not only for their being but also for their definition. And because everything that has matter is mobile,[c,7] it follows that *mobile being* is the subject of natural philosophy,[d,8] for natural philosophy concerns natural beings. Natural beings, however, are those whose principle is nature, and nature is the principle of motion and of rest in that in which it is. *Natural philosophy, therefore, concerns those things that have in them a principle of motion.*[e]

---

[a]"secundum rationem."

[b]Cf. John of St. Thomas, *Curs. Phil.*, Vol. I, qu. 27, art. 1, p. 825, col. a; *ibid.*, p. 826, col. b, 1. 26.

[c]Cf. St. Thomas, *In II Metaph.*, less. 4, n. 328, *In VIII Metaph.*, less. 1, n. 1686.

[d]Cf. St. Thomas, *In VI Metaph.*, less. 1, nn. 1155–1158; James A. McWilliams, S.J., *Physics and Philosophy* (Washington: Catholic University of America, 1936), pp. 1–27.

[e]Cf. St. Thomas, *In XI Metaph.*, less. 7, n. 2255; *De Trin.*, qu. 5, art. 1, corp.

4. Just as among all sciences *first philosophy*, wherein there is a determination of what is common to being as such, precedes,[9] so also in natural philosophy one treatise had to precede which would treat what is consequent upon mobile being in general. The reason for this is that what follow upon something common must be determined first of all and by themselves, so that we will not have to repeat them many times while treating all parts of that universal. And this is the book[a] of the *Physics*, which is also called *On Physical* or *Natural Hearing*[10] (because it was presented to listeners in the manner of teaching). Its subject is *mobile being and just that*.[b,11] I do not say *mobile body*, however, because in this treatise every mobile being is proved to be body, and no science proves its own subject. Thus, the treatise *On the Heavens*, which follows this one begins immediately with the notion of body. After this present treatise there are other books of natural philosophy which treat the species of mobile being. For example, in the *On the Heavens*, there is the treatment of being according to local motion, which is the first species of motion. *On Generation*, however, considers movement toward form and also considers first mobile beings, viz., the elements, in the light of their changes in general. Their specific changes are considered in *Of Meteors*. Mobile being that is composite and yet inanimate is studied in *On Minerals*, and animate being is treated in *On the Soul* and the books following it.[c]

---

[a]Actually, of course, there are eight books of this work.

[b]In this work we shall translate *"simpliciter"* by "absolutely," unless otherwise indicated, as here.

[c]Cf. Raymond Nogar, O.P., "Towards a Physical Theory," *The New Scholasticism*, Vol. XXV, No. 4 (Oct., 1951), pp. 397–438; William H. Kane, O.P., "The Extent of Natural Philosophy," *ibid.*, Vol. XXXI, No. 1 (Jan., 1957), pp. 85–97; Leo A. Foley, S.M., "The Persistence of Aristotelian Physical

5. Further, the Philosopher has set forth a preface for this book, and in it he shows the order of procedure in natural philosophy. He proceeds in two ways. First, he shows that we must begin from a consideration of principles, and secondly, he shows that among principles we must begin with the more universal ones (2). First, he presents the following argument: In all sciences which have principles, causes or elements, understanding and scientific knowledge proceed from the knowledge of principles, of causes, and of elements. But a science which concerns nature has principles, elements, and causes. Here, therefore, we must begin by determining principles.

When he says *to understand*, he refers to definitions, but when he says *to know*, he refers to demonstrations,[12] for, just as demonstrations are from causes, so too are definitions, since a complete definition is a demonstration differing in position alone,[13] as is said in the *Posterior Analytics*.[a]

But when he says *principles* or *causes* or *elements*, he does not intend to mean the same thing.[14] A cause is more than an element, for an element is that of which a thing is first composed, and which is in the thing, as is said in *Metaphysics* V,[b,15] just as letters of the alphabet, but not syllables, are elements of speaking. But that is called a cause which some things depend upon for their being or becoming. And so, even what is extrinsic to something or what is intrinsic to it, even though it is not the primary constituent of the thing, can be called a cause, but

---

Method," *The New Scholasticism*, Vol. XXVII, No. 2 (April, 1953), pp. 160–175; Charles de Koninck, "Les sciences expérimentales sont-elles distinctes de la philosophie de la nature?," *Culture*, Vol. II, No. 4 (Dec., 1941), pp. 465–476.

[a]Less. 16, n. 5.

[b]Less. 4, entire less., nn. 795–807.

not an element. A principle, on the other hand, implies a certain order of procedure, with the result that something can be a principle which is not a cause,[a] just as that from which motion begins is the principle of motion but not the cause, and as a point is the principle of a line but not the cause.

And so, by *principles* he seems to mean moving and agent causes,[16] in which the order of a certain procedure is especially observed. By the *causes* he seems to mean formal and final causes that things depend upon most for their being and becoming. By *elements* he seems to mean what are properly the first material causes. He uses these names disjunctively, however, and not copulatively to indicate that not every science demonstrates through all the causes.[b,17] Mathematics, indeed, demonstrates through the formal cause only. Metaphysics demonstrates through the formal and above all through the final cause and also through the agent cause. Natural philosophy, however, demonstrates through all the causes.

Moreover, the first proposition of the foregoing argument he proves from general opinion, as is done in the *Posterior Analytics*;[c] for we think that we know something when we know all its causes from the first to the last. And we should not take causes, elements, and principles other than as above, as the Commentator[d] would have it, but in the same way. Furthermore, he says *up to the elements*,

---

[a]Cf. St. Thomas, *In IV Metaph.*, less. 2, n. 548; *ibid.*, Bk. V, less. 1, n. 751.

[b]Cf. St. Thomas, *In I Post. Anal.*, less. 9, n. 5.

[c]Bk. I, less. 4.

[d]Averroes.

because what is last in cognition is matter,[18] since matter is for the sake of form, but form comes from the agent and is for an end, unless it be the end itself. For example, we say that a saw has teeth to cut and that those teeth must be of iron in order to be suitable for cutting.

6. Then, when he says: (2), he shows that among principles we must determine more universal ones first.[19] He shows this first by an argument, and secondly, by some examples (3). With regard to the first, he proposes the following argument: In knowing it is natural for us to proceed from what is better known to us to what is by nature better known.[20] But what is better known to us is confused, as universals are. Our procedure, therefore, should be from univerals to singulars.[a,21]

7. Further, to make the first proposition evident, he brings in the fact that to be better known to us is not the same thing as to be by nature better known. What is by nature better known is less known to us. And because for us the natural method or order of learning is to go from what we know to what we do not

---

[a]Cf. St. Thomas, *Summa Theol.*, Ia, qu. 85, art. 1, ans. 1; *In I Post. Anal.*, less. 4, nn. 15–16; *In Ethicorum ad Nichomachum*, Bk. I, less. 4, n. 52; *Scriptum Super Libros Sententiarum, Magistri Petri Lombardi*, Bk. II, dist. 3, qu. 3, art. 2, ans. 1. Here the "universale in re," the "universale a re" and the "universals ad rem" are distinguished. Cf. also *ibid.*, ans. 2–4; *De. Ver.*, qu. 8, art. 10, ans. 1. Here the universality from the viewpoint of the thing known and the universality from the viewpoint of the modes of knowing are distinguished. Cf. also *Summa Contra Gentes*, Bk. II, ch. 98. This studies universals from the viewpoint of modes of knowing. Cf. also *In Librum de Causis*, less. 10, nn. 247—253. Cf. Vincent Smith, "Abstraction and the Empiriological Method," *Proceedings of the American Catholic Philosophical Association*, Vol. XXVI (1952), pp. 35–50.

know,[a,22] we must go from what is better known to us to what is by nature better known.

We should note, however, that he says being known in nature is the same as being known absolutely. Those things are better known absolutely which are better known in themselves. But the things that are better known in themselves have more being, because a thing is knowable to the extent that it is a being. But those beings are to a greater extent[b] which are more in act. And so they are by nature more knowable. From our point of view, however, the opposite is true because in knowing, we proceed from potency to act. And our knowledge begins from sensible objects, which are material and potentially intelligible. And so, those things are better known to us than are separated substances, which are by nature better known, as is clear in *Metaphysics* II.[c] He does not say, therefore, *by nature better known*, as if nature knew them, but because they are better known in themselves and by their very nature. But he says *more known and more certain* because in the sciences not any kind of knowledge is sought but certain knowledge.[d,23]

---

[a]Cf. St. Thomas, *In I Post. Anal.*, less. 1, n. 9; *De Ver.*, qu. 11, art. 1, ans. 3.

[b]"magic."

[c]Less. 1, nn. 281–282.

[d]Cf. St. Thomas, *In I Post. Anal.*, less. 14; Charles De Koninck, "Introduction to the Study of the Soul," *Laval théol. et phil.*, Vol. III, No. 1 (1947), pp. 9–66. On the subject of probable vs. certain knowledge cf. St. Thomas, *In IV Metaph.*, nn. 572-577; *In I Post. Anal.*, less. 25; S. Edmund Dolan, S.F.C., "Resolution and Composition in Practical Discourse," *Laval théol. et phil.*, Vol. VI, No. 1 (1950), pp. 9–48.

To understand the second proposition we must know that here those things are called *confused* which have some potential and indistinct contents.[a] And because to know something indistinctly is intermediate between pure potency and perfect act, therefore, since our intellect proceeds from potency to act, the confused comes to it before the distinct. Complete knowledge, however, is actual when it is arrived at by resolution to the distinct knowledge of principles and elements. And this is why we know the confused before the distinct. But that universals are confused is obvious from the fact that universals contain their species in potency, and that whoever knows something in a universal way knows it indistinctly, whereas his knowledge becomes distinct when any one of those things that are contained potentially in a universal is actually known, for whoever knows animal knows rational only potentially. But it so happens that we know something in potency before we know it in act. In this order of learning, in which we proceed from potency to act, we therefore, happen to know animal before man.

8. This, however, seems to be contrary to what the Philosopher says in the *Posterior Analytics* I,[b,24] viz., that singulars are better known to us, but that universals are better known in nature or absolutely. But we should know that there he takes singulars as sensible individuals, and these are better known to us because for us sense cognition, which is that of singulars,[c] precedes intellectual cognition, which is that of universals. But because intellectual cognition is more perfect, and universals are intelligible in act, while singulars are not intelligible in act (since

---

[a]Cf. St. Thomas, *Contra Gent.*, Bk. II, chap. 96–100.

[b]Less. 2, entire lesson; *ibid.*, less. 4, nn. 15–16.

[c]Individuals.

they are material), universals are better known absolutely and according to nature. But here he takes singulars not as individuals but as species, and these are by nature better known, since they have more perfect being and distinct knowledge. Genera, however, are first known to us after the manner of potential and confused knowledge.

We should also note that the Commentator explains this in another way. Averroes says that the Philosopher in (2) wants to show the method of demonstrating belonging to this science, i.e., that it demonstrates through effects and through what is last in the order of nature. And so, what is said there refers to the procedure in demonstrating, and not in determining. When he says: (2b), however, he intends to show, he tells us, what is more known to us and by nature less known, i.e., what is composed in contrast to what is simple, understanding composed by the word *confused*.[a,25] Finally, he concludes that we should proceed from the more universal to the less universal, as if this were a kind of corollary.[b] And so, it is clear that the exposition of Averroes is not suitable because he does not bring the whole together under one intention, and because here the Philosopher does not intend to show the method of demonstration belonging to this science, for he will do this in the second book in accordance with the order of determination, and also because the confused ought not to be expressed as the composed, but rather as what is not distinct. Indeed, we cannot conclude anything from universals, since genera are not composed of species.

---

[a]Cf. St. Thomas, *De Unitate Intellectus Contra Averroistas*, chap. 6–7; Beatrice Zedler, "Averrhoes on the Possible Intellect," *Proceedings of the American Catholic Philosophical Association*, Vol. XXV (1951), pp. 164–178.

[b]Cf. St. Thomas, *In I Metaph.*, less. 2, n. 46.

9. Then, when he says: (3), he clarifies his point by three examples. The first of these is taken from the integral, sensible whole, and he says that the sensible whole is better known sensibly. The intelligible whole, therefore, is better known intellectually. But a universal is a kind of intelligible whole, because it comprises[a] many things as parts, i.e., its inferiors. For us, therefore, the universal is better known intellectually.

But this proof does not seem convincing, because he is using *whole*, *part*, and *comprehension* equivocally. But we should say rather that the integral whole and the universal whole agree in this, viz., that each is confused and not distinct, for just as whoever apprehends a genus does not apprehend the species distinctly but only potentially, so, too, whoever apprehends a house does not yet distinguish its parts. And so, since by reason of its confusion the whole is better known to us, this same reasoning is true of every whole. But to be a composite is not common to every whole. And so, it is clear that above[b] Aristotle deliberately said *confused* and not *composite*.

10. Then, when he says: (4), he gives another example regarding the integral, intelligible whole. Indeed, the thing defined is related to the defining elements somewhat as is an integral whole, i.e., to the extent that given the defining elements a thing is defined actually.[c] Still, whoever apprehends a word, e.g., that of *man* or *circle*, does not immediately distinguish the principal defining elements.

---

[a]Includes in its comprehension. This is the comprehension to which he is referring in the first sentence in the next paragraph.

[b]N. 7, par. 2, and par. 8, last part.

[c]"actu."

And so, the word is as a certain whole and indistinct thing, but the definition *divides into singulars*, i.e., it distinctly states the principles of the thing defined.

Now, this seems to be contrary to what he said above,[a,26] for the defining elements, which he said are first known to us, would seem to be the more universal.[b] Likewise, if the thing defined were more known to us than the defining elements, the thing defined would not be made known to us by the definition, for nothing is made known to us except from what is better known to us. Still we must say that we know the defining elements in themselves before we know the thing defined. But we know the thing defined before we know that such are its defining elements, just as we know animal and rational before we know man. We know man, however, in a confused way before we know that animal and rational are its defining elements.

11. Then, when he says: (5), he gives a third example, one taken from a sensible object under a more universal aspect. For, just as for us the more universal intelligible has primacy in intellectual cognition, e.g., animal before man, so also for us the more common sensible has primacy in sensible cognition, e.g., this animal before this man. And I say that it is sensibly prior both in place and in time. It is prior in place, because when we see someone from a distance, we see that it is a body before we see that it is an animal, and we see this before we see that it is a man, and lastly, we see that it is Socrates. Similarly, it is prior in time, for a child is aware of this as some man or other before he is aware of him as this

---

[a]N. 8.

[b]Cf. John of St. Thomas, *Curs. Phil.*, Vol. II, qu. 1, art. 3, col. a, 11. 4–33, p. 31.

man who is Plato, who is his father. And he says that *children first call all men fathers and all women mothers, but later they distinguish*, i.e., they recognize *each one* determinately. From this it is clearly shown that we know something confusedly before we know it distinctly.[a]

[a]Cf. Vincent Smith, *The General Science of Nature* (Milwaukee: The Bruce Publishing Company, 1958), pp. 33–38.

TEXT OF ARISTOTLE LESSON TWO[a]

1. Well, then, there must be either one principle of Nature or more than one. And if only one, it must be either rigid, as Parmenides and Melissus say, or modifiable, as the Physicists say, some declaring air to be the first principle, and others water. If, on the other hand, there are more principles than one, they must be either limited or unlimited in number. And if limited, though more than one, they must be two or three or four, or some other definite number. And if they are unlimited, they must either be, as Democritus held, all of the same kind generically, though differing in shape and sub-characteristics, or of contrasted nature as well.

2. The thinkers who inquire into the number of 'absolute entities,' again, follow the same line. For their first question is whether the constituents of which things are composed are one or more than one; and, if more than one, are they limited or unlimited? So they, too, are inquiring whether there is one principle or ultimate constituent, or many.

3. Now, as to the contention that all existence is one and is rigidly unchanging, we might say that it does not really concern the student of Nature, and he need not investigate it; any more than it concerns the geometer to argue with one who denies the geometrical axioms. Such questions must be dealt with either by some other special science or by a fundamental discipline that underlies all the sciences. And so it is in this matter of the unity of the natural principle, for if it is only one, and one in the sense of rigidity, it is not a principle at all; for a principle must be the principle of something or things other than its naked self.

4a. To consider, therefore, whether the natural principle is one in this sense, is like discussing any other paradox that is set up just for the sake of arguing, like the paradox of Heracleitus,

---

[a]Includes 11. 184 b 15—185 a 20, Wicksteed and Cornford, *op. cit.*, pp. 14–19.

4b. or the contention (should anyone advance it) that the totality of existence is one single man. If their fundamental paradox is admitted, there is nothing strange in all their other paradoxes following from it.

5. Let us then start from the datum that things of Nature, or (to put it at the lowest) some of them, do move and change, as is patent to observation; and let us make a note that we are not bound to answer every kind of objection we may meet, but only such as are erroneously deduced from the accepted principles of the science in question. Thus, it is the geometer's business to refute the squaring of the circle that proceeds by way of equating the segments, but he need not consider Antiphon's solution.

6. And yet, after all, though their (the Eleatics') contention does not concern the study of Nature, it does incidentally raise points that are of interest to the physicist; so perhaps it will be as well to examine it briefly,—especially as it has a certain philosophical interest of its own.

The opinions of the ancient philosophers regarding the principles of nature and of being. It is not within the province of natural philosophy to refute the opinions of some of these.

12(1). After the preface in which it was shown that the philosophy of nature ought to begin from more universal principles, he begins, *according to the order stated above, by discussing what pertains to natural philosophy*. And he does this in two parts. In the first part, he studies the universal principles of natural philosophy. In the second part, he studies mobile being in general, which he intends to study in this book and in the third book (less. 1, n. 1). The first part is divided again into two parts. In the first, he studies the principles of the subject of this science, i.e., the principles[a] of mobile being as such.[27] He considers the principles of this doctrine in the second part, i.e., in the second book (less. 1, n. 1).[28] The first of these last two divisions is again divided into two. First, he discusses the opinions of others regarding the common principles of mobile being, and in the second part, he investigates the truth with regard to these principles (less. 10, n. 1).

In reference to the first, viz., the division of the opinions, he proceeds in a threefold manner. First, he states the various opinions of the ancient philosophers

---

[a]On the meaning of "principles" here, cf. John of St. Thomas, *Curs. Phil.*, Vol. II, p. 5, col. a, 1. 18 ff.

concerning the common principles of nature; secondly, he shows that to discuss some of these opinions is not within the province of natural philosophy (3); thirdly, he continues the discussion of their opinions in arguing against their falsity (less. 3, n. 1). With regard to the first, viz., the stating of the various opinions of the ancient philosophers, he proceeds in a twofold manner. First, he states the different opinions of philosophers regarding the principles of nature; secondly, he shows that there is a common difference in the opinions of the philosophers about beings (2).

13(2). He says first, therefore, that there must be either one principle of nature or many. And each viewpoint had philosophers who gave opinions in support of it.

For some of them posited one principle, and some many. And of those who stated one, some said that it was immobile, as Parmenides and Melissus, whose opinions will be clear below.[a] Yet, some, viz., the ancient natural philosophers, held that it was mobile. Of these, some, e.g., Diogenes,[b] stated air to be the principle of all natural things; others, e.g., Thales,[c] held that it was water, and still

---

[a]Less. 3, nn. 5–6.

[b]Frag. 7: "And this (Element) itself is a body both everlasting and immortal; whereas of other things, some come into being and others pass away." Kathleen Freeman, *Ancilla to the Pre-Socratic Philosophers* (Oxford: Basil Blackwell, 1956), p. 90. In this and subsequent citations from Freeman, all parentheses include words that she has added for the purpose of clarification, or summaries of long quotes, unless otherwise specified.

[c]"Thales of Miletus was in his prime about 585 B.C. Whether he ever wrote a book is unknown; if he did, no genuine fragment survives." *Ibid.*, p. 18.

others, e.g., Heraclitus[a] held that it was fire; yet, there were those who held that it was some mean between air and water, as vapor. But none of those who stated only one principle said that it was earth, on account of its bulkiness.[b] But they stated mobile principles of this kind because they said that other things were made by the rarefaction and condensation of any one of these.

But of those who posited many principles, some held that they were finite, and some that they were infinite.[29] Of those who stated that they were finite, although more than one, some held that there were two, viz., fire and earth, as he will say below[c] about Parmenides, and some said three, viz., fire, air and water, for they considered the earth as a composite on account of its bulkiness. Others,

---

[a]"Heraclitus of Ephesus was in his prime about 500 B.C. He wrote one book, covering all knowledge, metaphysical, scientific and political, in an oracular style." *Ibid.*, p. 24; frag. 30: "This ordered universe (cosmos), which is the same for all, was not created by any one of the gods or of mankind, but it was ever and is and shall be ever-living Fire, kindled in measure and quenched in measure." *Ibid.*, p. 26; frag. 31: "The changes of fire: first, sea; and of sea, half is earth and half fiery waterspout....Earth is liquified into sea, and retains its measure according to the same Law as existed before it became earth." *Ibid.*, p. 27.

[b]Cf. Xenophanes of Colophon who, though he stresses earth as a principle also adds water and air as principles. According to Freeman, Xenophanes "was in his prime about 530 B.C." *Ibid.*, p. 20; frag. 27: "For everything comes from earth and everything goes back to earth at last." *Ibid.*, p. 23; frag. 28: "This is the upper limit of earth that we see at our feet, in contact with the air; but the part beneath goes down to infinity." *Ibid.*; frag. 29: "All things that come into being and grow are earth and water." *Ibid.*; frag. 30: "The sea is the source of water, and the source of wind. For neither could (the force of the wind blowing outwards from within come into being) without the great main (sea), nor the streams of rivers, nor the showery water of the sky; but the mighty main is the begetter of clouds and winds and rivers." *Ibid.*; frag. 33: "We all have our origin from earth and water." *Ibid.*, p. 24.

[c]Less. 10, n. 2.

indeed, e.g., Empedocles,[a] held that there were four, or even according to some, another number,[b] for even Empedocles himself added two others with the four elements, viz., friendship and strife.

Those who held many infinite principles differed. Democritus, indeed, stated that individual bodies, which are called atoms,[c] were the principles of all things. But he said that bodies of this kind were, in their nature, all of one genus, but that they still differed in figure and shape,[d] and not only differed, but had mutual contrariety.[30] For he stated three contrarieties, one in figure, which is curved and straight, the second in order, which is prior and posterior, and the third in position, viz., before and behind, up and down, right and left. And so, from these existing bodies of one nature, he said that different things were made according to a difference of figure, position and order of atoms. But from this opinion Aristotle proceeds to mention an opposite one, i.e., that of Anaxagoras who posited infinite

---

[a]"Empedocles of Acragas was in his prime about 450 B.C. He wrote two poems in hexameter verses...." Freeman, *op. cit.*, p. 51.

[b]Cf. also Philolaus of Tarentum of the fifth century, frag. 12: "The bodies (physical Elements) of the Sphere are five: the Fire in the Sphere, and the Water, and Earth, and Air, and fifth, the vehicle (?) of the Sphere." *Ibid.*, p. 75. With regard to the way in which the ancient philosophers posited a different number of principles, cf. John Burnet, *Early Greek Philosophy* (London: Adam and Charles Black, 1948).

[c]"Democritus of Abdera was in his prime about 420 B.C." Freeman, *op. cit.*, p. 91; frag. 9: "Sweet exists by convention, bitter by convention, colour by convention; atoms and Void (alone) exist in reality....We know nothing accurately in reality, but (only) as it changes according to the bodily condition, and the constitution of those things that flow upon (the body) and impinge upon it." *Ibid.*, p. 93.

[d]Cf. St. Thomas, *infra*, Bk. VII, less. 5, nn. 2, 5.

principles. But these, by their nature, were not of the same genus.[a] For he considered the principles of nature to be infinite minimal parts of flesh or bone, and of other materials of this kind, as will be clear further below.[b]

In addition, we should observe that he does not divide the many principles into mobile and immobile. The reason for this is that no one who posits many first principles can posit them as immobile for, since all posit contrariety in principles, and since contraries are by nature opposed to one another, immobility is inconsistent with a plurality of principles.

14(3). Then, when he says: (2), he shows that there is a similar difference of opinion regarding beings. And he says that, likewise, natural philosophers, inquiring about things that are, i.e., about beings,[31] want to know how many beings there are, i.e., whether one or many, and, if they are many, they want to know whether they are finite or infinite. And, the reason for this is that ancient philosophers knew only the material cause,[c] and hardly touched upon the other causes.[32] Besides, they said that natural forms were accidents, just as artificial ones are.[33] And thus, just as the whole substance of artificial entities is their matter, so also, according to them, it followed that matter was the whole substance of natural

---

[a]According to Freeman, *op. cit.*, p. 82, Anaxagoras of Clazomenae wrote one book and "was in his prime about 460 B.C." Cf. frag. 1: "(Opening sentences from his book 'On Natural Science'): All things were together, infinite in number and in smallness. For the Small also was infinite. And since all were together, nothing was distinguishable because of its smallness. For Air and Aether dominated all things, both of them being infinite. For these are the most important (Elements) in the total mixture, both in number and in size." *Ibid.*, p. 83.

[b]*Infra.*, less. 9; *ibid.*, less. 11, nn. 77–81.

[c]St. Thomas, *In I Metaph.*, less. 4, nn. 74 ff.

beings.[34] Thus, those who posited only one principle, e.g., air, thought that other beings were air in substance, and the same is true with regard to the other opinions. And this is what Aristotle means when he says that natural philosophers seek *the constituent elements of existing things*, i.e., in seeking the principles they seek out the material causes of beings. Thus, it is clear that when they investigate whether beings are one or many, their inquiry concerns material principles. These are called elements.

15(4). Then, when he says: (3), he shows that it is not the task of natural philosophy to refute some of their opinions. And he treats this in two ways. First, he shows that it is not within the province of natural philosophy to refute the opinions of Parmenides[a] and Melissus.[b] Secondly, he gives a reason why, for the present, it is useful to refute them (6). He treats the first in two ways. First, he shows that it is not the task of natural philosophy to refute the foregoing opinion. Secondly, he shows that it is not the task of natural philosophy to resolve the arguments which are brought forth for proving this opinion (4b). He shows this by two reasons. The second of these reasons begins at (4b).

First, he says, therefore, that it is not the function of natural philosophy to devote itself to examining this opinion, viz., whether being is one and immobile.

---

[a]"Parmenides of Elea was in his prime about 475 B.C. He wrote a poem in hexameter verse, addressed to his pupil Zeno; it was divided into three parts: the Prologue, the Way of Truth, the Way of Opinion." Freeman, *op. cit.*, p. 41.

[b]"Melissus of Samos was in his prime about 440 B.C. He wrote a treatise *On Being* in defense of Parmenides' theory." *Ibid.*, p. 48.

For it has already been shown[a] that, according to the meaning[b] of the ancient philosophers, positing one immobile principle is the same as stating one immobile being.

That it is not within the scope of natural philosophy to refute this opinion he shows in this way: It is not the task of geometry to offer an argument against someone who denies the principles of geometry. But this either belongs to some other particular science (if geometry is subordinated to some other particular science, just as music is subordinate to arithmetic,[c] and it is in the province of arithmetic to refute the position of one who denies the principles of music), or this pertains to a general science, i.e., to logic or to metaphysics. But the foregoing position denies the principles of nature, because, if there is only one being, and if it is *one in such a way*, viz., an immobile one, so that other beings cannot come to

---

[a]N. 3.

[b]"intentio."

[c]Cf. the ff. very interesting fragment of Archytas of Tarentum from the fourth century B.C. and who "wrote in literary Doric a work on *Mathematical Science*, and on *Harmony*; possibly also on *Mechanics*." Frag. 1: "Mathematicians seem to me to have excellent discernment, and it is in no way strange that they should think correctly concerning the nature of particular existences. For science they have passed an excellent judgment on the nature of the Whole, they were bound to have an excellent view of separate things. Indeed, they have handed on to us a clear judgment on the speed of the constellations and their rising and setting, as well as on (plane) geometry and Numbers (arithmetic) and solid geometry, and not least on music; for these mathematical studies appear to be related. For they are concerned with things that are related, namely the two primary forms of being.

First of all therefore, mathematicians have judged that sound is impossible unless there occurs a striking of objects against one another. This striking, they said, occurs when moving objects meet one another and collide...." Freeman, *op. cit.*, p. 78.

be from it, the nature of principle is taken away, because every principle is either the principle of some one thing or of several things. Multitude, therefore, follows upon the positing of a principle, since one thing is a principle and that of which it is a principle is another. Whoever, then, denies multitude denies principles. And so, it is not for a natural philosopher to argue against this position.

16(5). Then, when he says: (4), he shows this same thing by another argument. No science, indeed, is required to offer argument against obviously false and improbable opinions, for it is absurd to be concerned if someone maintains an opinion contrary to those of a wise man, as is said in *Topics I.*[a] This, therefore, is what he means, viz., that to attempt to inquire accurately if being is one in this manner, viz., immobile, is the same as refuting any other improbable position, as, for example, the position of Heraclitus, who said that all things were always being moved[b] and that nothing was true, or the position of one who would say that all being is one man. This view, indeed, would be entirely improbable. But whoever maintains that there is only one immobile being, is forced to maintain that all being is some one thing. It is evident, therefore, that it is not the function of natural philosophy to argue against this position.

17(6). Then, when he says: (4b), he shows, moreover, that it is not within the province of natural philosophy to resolve the arguments of the philosophers

---

[a]St. Thomas, Bk. VIII, less. 5.

[b]Cf. Heraclitus, frag. 90: "There is an exchange: all things for Fire and Fire for all things, like goods for gold and gold for goods." Freeman, *op. cit.*, p. 31; cf. also frag. 91: "It is not possible to step twice into the same river." *Ibid.*

mentioned above. And he gives two reasons for this. The second of these is stated at (5). And so, he first proves the point by the fact that no science is required to resolve sophistic arguments, which have an obvious defect either of form or of matter. And this is what he means, viz., that paying attention to improbable arguments is similar to *resolving a contentious argument*, i.e., a sophistic argument.[a] Now the arguments both of Melissus and Parmenides are sophistic, for the content is false.[b] And so, he says that *they begin with false assumptions*, i.e., they assume as true propositions that are false. And they are defective in form, and so, he says that *they are not 'syllogizings.'* But the argument of Melissus is *more troublesome*, i.e., empty and futile, and it *is not defective*, i.e., it does not cause doubt; and this is shown below.[c] It is not illogical, however, if, when one absurdity is presented, that others follow. We can, therefore, conclude that the natural philosopher is not required to solve arguments of this nature.

18(7). Then, when he says: (5), he gives a second argument for this, and it is as follows: In natural philosophy, it is supposed that *either all or some* natural beings are moved. He says this because with regard to some beings, e.g., the soul, the center of the earth, the poles of the sky, natural forms, and other beings of this nature, it is doubtful if they are moved and how they are moved. And that natural beings are moved can be shown from induction, because it is sensibly apparent that

---

[a]Cf. Aristotle, *Topica*, Bk. I, chap. 1.

[b]Their arguments are materially defective or erroneous.

[c]Less. 3, 5.

natural beings are moved.[a,35] Besides, motion must be supposed in natural philosophy, just as nature must be supposed. Motion is stated in the definition of nature, for nature is the principle of motion, as will be said below.[b] In addition, granted that natural philosophy supposes motion, he proceeds further to the point to show from this that not all arguments are to be resolved in any one science but only those arguments which conclude falsely from the principles of that science. Those arguments, however, that do not conclude from the principles of a science but rather from the contraries of its principles are not resolved in that science.

And he proves this from an example in geometry which seeks to disprove a *tetragon*, i.e., the squaring of a circle. To accomplish *such a squaring through divisions* of the circumference would belong to geometry, because it supposes nothing contrary to the principles of the science. For someone wanted to find a square equal to a circle by dividing the circumference of the circle into many parts, and by drawing straight lines to the single parts. And in this way, by discovering some figure, viz., a rectilinear equal to some one of those figures which are contained by the cutting of the circumference and a cord, either to many or all of those figures, he thought that he had found a rectilinear figure equal to the whole circle, and that by the principles of geometry it was easy to find a square equal to a circle. And so, he thought that he could find a square equal to a circle.[c] But the argument was insufficient, because, although those divisions would consume the whole circumference of the circle, yet, the figures contained by the cutting of the circumference and by the straight lines did not comprise the whole circular surface.

---

[a]Cf. St. Thomas, *In IV Metaph.*, less. 12, nn. 682–683.

[b]Bk. II, less. 1, n. 4 (last part); Bk. III, less. 1, n. 2.

[c]Cf. *infra.*, Bk. VIII, less. 5, n. 3.

But to disprove the square of Antiphon,[a] is not proper to geometry, because he used the contraries of the principles of geometry, for he inscribed in a circle a certain rectilinear figure, e.g., a square, and he divided each arc into two halves, which the sides of the square subtended, and from the points of the divisions he drew a straight line to all corners of the square. And in this way, there resulted in the circle a figure of eight angles, and this figure conformed more to the equality of the circle than did the square. Again, he divided into two halves each of the arcs, which the sides of the octangular figure subtended. Thus, by drawing straight lines from points of the divisions to the angles of the above-mentioned figure, a figure of sixteen angles resulted. And this conformed even more to the equality of the circle. By always dividing the arcs, therefore, and by drawing straight lines to the corners of the pre-existing figure, he obtains a figure more nearly disposed to the quality of the circle.

Antiphon says, moreover, that it is not possible to proceed to infinity[b] in the division of the arcs, and so, one must arrive at some rectilinear figure equal to a circle to which one can equate a square. Since he supposed that arcs are not always divided into halves, which is contrary to the principles of geometry, therefore, it is not the function of geometry to refute an argument of this kind. Because the arguments of Parmenides and Melissus suppose that being is immobile, as will be explained below,[c] and since this is against the principles supposed in natural

---

[a]"Antiphon the Sophist, believed to be of Athens, and to have lived in the latter half of the fifth century B.C." Freeman, *op. cit.*, p. 144; cf. frag. 13: "(Aristotle, Physics 185a: Antiphon's construction for the squaring of the circle by means of the inscription of triangles)." Freeman, *ibid.*, p. 145.

[b]Cf. St. Thomas, *infra*, Bk. III, less. 9.

[c]Less. 2 and 5.

philosophy, it follows, therefore, that it is not the function of natural philosophy[36] to resolve arguments of this kind.

19(8). Then, when he says: (6), he gives the reason why he argues against the foregoing position. He says that because the philosophers mentioned above spoke about natural beings, even though they themselves did not bring in the actual *fallacies*, i.e., difficulties in the order of natural philosophy, it is useful for our purpose to argue against opinions of this kind. The reason for this is that although it is not the task of natural philosophy to argue against such positions, yet, it does belong to first philosophy.[a,37]

---

[a]Cf. St. Thomas, *In I Metaph.*, less. 4, n. 78; *In IV Metaph.*, less. 5, n. 593; *In VI Metaph.*, less. 1, n. 1170; *In XI Metaph.*, less. 7, n. 2267.

TEXT OF ARISTOTLE LESSON THREE[a]

1. Now since the term 'existent' is itself ambiguous, it lies at the very heart of the matter to inquire whether (1) they who assert all existing things to be 'one' are thinking of all existing things substantively or quantitatively or qualitatively. Or (2) are all existing things 'one substance,' like 'one man' or 'one horse' or 'one soul'? Or is it that they are all one in quality, (say) all the same 'white,' or 'hot,' or so forth, so that the same qualitative predication can be made of them all? For though all such assertions are impossible, yet they are far from being all identical.

(1) Thus (a) if all things are to be one both substantively, quantitively, and qualitatively, then (whether we consider these forms of 'being' as objectively separable from each other or not) in any case existences are many and not only one. Whereas (b) if it be meant that all things are one *magnitude*, or one *quality*, then, whether there be any substantive existence at all or no, the assertion is absurd—if we may so call the impossible. For none of the categories except 'substance' can exist independently, since all the rest must necessarily be predicated of some substance as their *subjectum*. But Melissus says that the Universe is unlimited, which would make it a magnitude or *quantum*, since 'limited' and 'unlimited' pertain to *quanta* only, so that no substantive being and no quality or affection can, as such, be called unlimited (though it may incidentally be a *quantum* also); for the conception of a *quantum* enters into the definition of 'unlimited,' whereas the conception of substantive existence, or quality, does not. If then (a) the 'existent' is both a substance and a magnitude, it is two and not one; whereas (b) if it is a substance only, it cannot be unlimited, nor indeed can it have any dimension at all, for otherwise it would be quantitative as well as substantive.

2(2). Again, since 'one' is itself quite as ambiguous a term as 'existent,' we must ask in what sense the 'All' is said to be *one*. (a) Continuity establishes one kind of unity; (b) indivisibility another; (c) identity of definition and of constituent characteristics yet a third; as for instance (the Greek) *methy* and *oinos* (both of which mean 'wine') are 'one and the same' thing.

---

[a]Includes 11. 185 a 21—185 b 25, Wicksteed and Cornford, *op. cit.*, pp. 19–25.

So then, (a) if a single *continuum* is what is meant by 'one,' it follows that 'the One' is many, for every *continuum* is divisible without limit. (And this suggests the question—not to our present purpose, perhaps, but interesting on its own account: whether the part and the whole of a *continuum* are to be regarded as a unity or as existing severally, and in either case in what sense they are a unity or plurality, and if they are regarded as a plurality in what sense a plurality. And the question arises again with respect to a discontinuous whole, consisting of unlike parts: In what sense do such parts exist, as several from the whole? Or if each is indivisibly one with the whole, are they so with each other?)

3. Whereas (b) if 'one' signifies an 'indivisible,' then it excludes quantity (as well as quality); and therefore, the One Being, not being a quantum, cannot be 'unlimited' as Melissus declares it to be; nor indeed 'limited' as Parmenides has it, for it is the limit only, not the *continuum* which it limits, that complies with the condition of indivisibility.

4. Lastly (c) if the contention is that all things are identically 'one and the same' by definition (as 'clothes' and 'garments' are) we are back again at the Heracleitean paradox; for, in that case, being good and being bad will be the same and being not-good the same as being good—with the consequence that the same thing will be both good and not good, or both a man and a horse, and we shall no longer be maintaining that all existences are one, so much as that none of them is anything—and being of a certain quality will be the same thing as being of a certain magnitude.

The refutation of the opinion of Parmenides and Melissus who said that all things are one being.

20(1). Having presented the opinions of the philosophers about principles, *he here argues against them*. First, he argues against those who do not speak about nature in the way a natural philosopher does (less. 8, n. 1), and secondly, against those who do speak about nature in the manner of a natural philosopher. With regard to the first, he proceeds in a twofold manner. First, he argues against what Melissus and Parmenides stated, and secondly, against their manner of argument (less. 5, n. 1). The first, viz., what they said, he treats in a twofold way. First, by an argument taken from the viewpoint of *being*, which is the subject in the proposition, he argues against the statement that *being is one*.[a] Secondly, he argues against that statement by an argument taken from the viewpoint of *one* which is the predicate (2).

21(2). First, he says, therefore, that what is above all to be taken as a principle in arguing against the foregoing position is that *what is*, i.e., being, has many significations.[38]

[a]Cf. Frederick Copleston, S. J., *History of Philosophy* (Westminster, Md.: The Newman Bookshop, 1946), Vol. I, Greece and Rome, pp. 47–53.

We must ask those who say that being is one what meaning they attach to the term being, i.e., whether they take it as substance, quality, or some other genus. And because substance is divided into universal and particular, i.e., into first and second substance,[a,39] and again into many species, we should ask whether they say that being is one as one man, one horse, one soul, or one quality, e.g., white, warm or something of this kind, for it makes a great deal of difference which of these is meant.

If, therefore, being is one, either it must be substance and accident at the same time, or it must be accident only, or substance only. If, however, it is substance and accident at the same time, it will not be only one being, but two. It does not make so much difference with regard to whether or not substance and accident are simultaneously in one thing as one or as diverse, because, even though they may be simultaneously in one thing, still, they are not one absolutely, but one in subject. And so, by positing substance with accident, it follows that they are not one absolutely but many. If, however, we say that it is accident only and not substance, this is entirely impossible, for accident can in no way be without substance, since all accidents are predicated of substance, as of a subject, and this is what accident means.

If, on the other hand, we say that it is substance only without accident, consequently, there is no quantity, for quantity is an accident, and this is against the position of Melissus. He, indeed, said that being is infinite,[b] and consequently, it

---

[a]Cf. St. Thomas, *Quaestiones Disputatae De Potentia Dei*, qu. 9, art. 2, ans. 6; *Summa Theol.*, Ia, qu. 29, art. 1, ans. 1–2; John of St. Thomas, *Curs. Phil.*, Vol. I, qu. 15, art. 2–3, pp. 530 ff.

[b]Cf. Melissus, frag. 3: "But as It is always, so also its size must always be infinite." Freeman, *op. cit.*, p. 48; frag. 4: "Nothing that has a beginning and an

is quantified, because the infinite, strictly speaking,[a] is only in quantity.[b] But substance and quality and such things are called infinite only accidentally, i.e., to the extent that they are simultaneously with quantity. Since, Melissus posits infinite being, he cannot, therefore, posit substance without quantity.[40] If, therefore, it is substance and quantity together, it follows that there is not only one being but two; but if it is substance alone, it is not infinite, because it will have neither magnitude nor quantity. What Melissus says, viz., that being is one, can, therefore, in no way be true.

22(3). Then, when he says: (2), he gives the second argument, which is taken from the viewpoint of unity. And he treats this in a twofold way. First, he presents his argument, and secondly, he shows how some erred in breaking it down (less. 4, n. 1). And so, he says first that just as being has many meanings, so also does the term one, and for this reason we should consider in what way they say that all things are one. For the term one is understood in three ways, either as the continuous is one,[41] e.g., a line or a body, or as the indivisible is one, e.g., a point, or as those things whose meaning or definition is one are called one, as vinegar and wine are called one. First, he shows that they cannot say, therefore, that all things are one by continuation, because the continuous is in a certain sense many, for whatever is continuous is divisible to infinity, and thus contains in itself

---

end is either everlasting or infinite." *Ibid.*

[a]"per se loquendo."

[b]On the infinite, cf.: St. Thomas, *In III Physic.*, less. 6-13; *Summa Theol.*, Ia, qu. 7, art. 2-4; *De Ver.*, qu. 2, art. 2, ans. 5; *ibid.*, art. 10.

many parts. And so, whoever states continuous being must admit that it is in some sense many.

And this occurs not only by reason of the multitude of parts, but also on account of the difference which there seems to be between the whole and the parts, for there is a doubt whether the whole and the parts are one or many. And although this doubt may not be pertinent for our purpose, yet, in itself it is useful to know. The same is true not only of continuous wholes but also of contiguous wholes whose parts are not continuous, e.g., parts of a house which are one in contact and composition. And it is clear that a whole is relatively[a] but not absolutely the same as a part. For, if the whole were absolutely the same as one of the parts, for the same reason it would be the same as another of the parts. But things that are equal to one and the same thing are mutually the same, and so, it follows that both parts, if they are stated to be absolutely the same as the whole, are mutually the same. And from this it would follow that the whole is indivisible, since it does not have a difference of parts.

23(4). Then, when he says: (3), he shows that all things cannot be one as the indivisible is one, because what is indivisible cannot be quantified, since all quantity is divisible, and consequently, it cannot be qualified, since quality should be understood to be founded upon quantity.[42] And, if it is not quantified, it cannot be finite, as Parmenides said, nor infinite, as Melissus said,[b] because an indivisible

---

[a]"secundum quid."

[b]Cf. Melissus, frag. 5: "If it were not One, it will form a boundary in relation to something else." Freeman, *op. cit.*, p. 48; cf. also frag. 6: "If it were infinite, it would be One; for if it were two, (these) could not be (spatially) infinite, but each would have boundaries in relation to each other." *Ibid.*

terminus, as a point, is an end and not finite, because the finite and the infinite[43] agree in this, viz., that each bespeaks quantity.

24(5). Then, when he says: (4), he shows why we cannot say that all things are one in meaning, because if this were the case, three absurdities would result. The first is that contraries would be one in meaning,[a,44] i.e., the meaning of good and of bad would be the same, as Heraclitus stated the meaning of contraries to be the same, as is clear in *Metaphysics* IV.[b,45] The second absurdity is that the meaning of good and not good would be the same, because the not-good follows upon evil, and so the meaning of being and non-being would consequently be the same.[c,46] And from this it would also follow that all beings would not be one being only, as they stated, but would also be non-being or nothing, because things that are one in meaning are related in such a way that whatever is predicated of the one may also be predicated of the other. Hence, if being and nothing are one in meaning, it follows, if all things are one being, that all things are nothing. The third absurdity is that different genera, e.g., quantity and quality, would be the same in meaning. And he states this absurdity when he says *for both quality and quantity*.

---

[a]Cf. *infra*, less. 15; St. Thomas, *In IV Metaph.*, less. 15, n. 719.

[b]Cf. less. 6, n. 606 and less. 17, n. 737. Cf. also Heraclitus, frag. 8: "That which is in opposition is in concert, and from things that differ comes the most beautiful harmony." Freeman, *op. cit.*, p. 25; frag. 51: "They do not understand how that which differs with itself is in agreement; harmony consists of opposing tension, like that of the bow and the lyre." *Ibid.*, p. 28; frag. 54: "The hidden harmony is stronger ((or) 'better') than the visible." *Ibid.*

[c]Cf. St. Thomas, *In I Sent.*, dist. 19, qu. 5, art. 1, ans. 8.

We should note, however, that as the Philosopher says in *Metaphysics* IV,[a,47] a demonstration strictly speaking, which proceeds from the more known absolutely, cannot be brought against those who deny principles, but only an *argumentum ad hominem*,[b] which proceeds from what is supposed by an opponent and which is sometimes less known absolutely.[48] And so, in this refutation, the Philosopher uses many things which are less known than this, viz., that beings are many and not one only, and he gives arguments for this.

---

[a]Less. 6.

[b]"demonstratio ad contradicendum."

TEXT OF ARISTOTLE

LESSON FOUR[a]

1. Indeed some of the later ancients were themselves disturbed by the danger of finding themselves admitting that the same thing is both one and many; and therefore some of them, like Lycophron, banned the word 'is' altogether; while others did such violence to language as to substitute for 'the man is pale-complexioned' the phrase 'the man has complexion-paled,' or to admit 'the man walks' but not 'the man is walking,' for fear of making the one existent appear many by affixing the word 'is'; which nervousness resulted from their equating 'is' with 'exists' and ignoring the different senses of 'one' or of 'being.' But the fact is that one thing obviously *can* be many, either in the sense of being several conceptually distinct things at once (for though it is one thing to be pale-complexioned and another to be cultivated, yet one and the same man can be both, so that 'one' is 'many'); or in the sense in which a whole can be regarded as being the sum of the parts into which it can be divided. This latter case brought our philosopher to a stand; and they had to admit multiplicity in unity, unwillingly, as though it were a paradox. And so, of course, it would be, if 'one' and 'many' were used in the sense in which they contradict each other, but not if, for instance, we are regarding the 'one' both in its actual unity and its potential multiplicity.

---

[a]Includes 11. 185 b 26—186 a 3, Wicksteed and Cornford, *op. cit.*, pp. 25–26.

How later philosophers also fell into the error of the preceding, i.e., that the one and the many can in no way agree.

25(1). After the Philosopher has disproved the opinion of Parmenides and Melissus who stated that being is one, *he shows that some subsequent philosophers fell into error for the same fundamental reason.*

Parmenides and Melissus, indeed, erred because they did not know how to distinguish the meaning of *one*. And so, what is one in some way they proclaimed to be one absolutely. Subsequent philosophers, who also did not know how to distinguish the meaning of one, considered it absurd that the same thing should in some way be both one and many. Nevertheless, convinced by arguments, they were forced to acknowledge this. And he says that subsequent philosophers *were thrown into confusion* by this, i.e., they fell into doubt, just as did the ancient philosophers, viz., Parmenides and Melissus, because they were afraid that they might be forced to say that the same thing is one and many, which seemed absurd to both men. And the first ones, therefore, who said that all things are one, did away with multitude entirely, but subsequent philosophers attempted to deny multitude to whatever they stated to be one.

26(2). And so, some, e.g., Lycophron,[a] in their propositions, removed the word *is*, for they said that we should not say *man is white*, but we should say *white man*, for they considered that man and white were in some sense one. Otherwise, white would not be predicated of man. But it seemed to them that this word *is*, since it is a verbal copula, would unite two things. And so, since they wanted to remove multitude entirely from what is one, they said that this word *is* must not be stated.

But because the expression seemed incomplete and because it seemed that an incomplete meaning[49] was begotten in the mind of the hearer if the terms were not conjoined by the addition of some verb, others, who wanted to correct this, changed this manner of speaking, and did not say *white man* on account of the imperfection of the expression. Nor did they say *man is white*, for fear that this might imply multitude, but they said *man whitens*,[b] because this, i.e., *to whiten*, seemed to them to imply nothing except some transmutation in a subject.[c,50] And similarly, they said that we should not say *man is walking* but *man walks*, for fear that by the addition of this verbal copula *is*, what they considered to be one, viz., *white man*,

---

[a]"Lycophron 'The Sophist': birthplace unknown; lived probably in the first half of the fourth century B.C. An orator of the school of Gorgias; interested also in metaphysics, political science and politics." Frag. 1: "Knowledge is an association between the act of knowing and the soul." Frag. 2: "(He eliminated the verb 'is' in predication)." Freeman, *op. cit.*, p. 139.

[b]"homo albatur."

[c]"singulariter."

they might make to be many, as if one and being were said *in one and the same sense*,[a] i.e., in one way and not in a multiple sense.[b,51]

27(3). But this is false, because what is one in one way can be many in another way, just as what is one in subject can be many in meaning, for there is one meaning for musical and another for white. We can conclude from this that one is many. In another way, it happens that what is one in its totality and in act may be many with regard to the division of the parts, and so, the whole is one in its totality but has a multitude of parts. And although they might discover some remedy for what is one in subject and many in meaning, like taking away the word *is* or changing the manner of speaking, as was said above,[c] yet, *here*, viz., with regard to the whole and the parts, they failed completely, since they did not know how to answer the problem, and they admitted that it is absurd that the one be many. But this is not absurd when one and many are taken as opposites, for one in act and many in act are opposed, but one in act and many in potency are not opposites. And so, he adds that *one* is said in a variety of ways, i.e., *one* in potency and one in act, and in this way nothing prevents the same thing from being one in act and many in potency, as is clear in the case of the whole and the parts.

---

[a]Cf. *infra*, Bk. III, less. 3, n. 2.

[b]On the problem of the noetic one and the many, cf.: St. Thomas, *De Ver.*, qu. 8, art. 18; *Contra Gent.*, Bk. II, chap. 96-100 incl.; *ibid.*, Bk. III, chap. 112; Charles De Koninck, "La dialectique des limites comme critique de la raison," *Laval théol. et phil.*, Vol. I, No. 1 (1945), pp. 177–185; Juvenal Lalor, O.F.M., "Notes on the Limit of a Variable," *ibid.*, pp. 129–150.

[c]N. 2.

28(4). Finally, however, he draws the conclusion that he principally intended, viz., that it is clear from the foregoing reasons that it is impossible that all beings be one.

1. Following up this line, we shall see that it is impossible for 'all things to be one,' and we shall have no difficulty in refuting the arguments by which the assertion that they are is supported. For both Parmenides and Melissus argue sophistically, inasmuch as they make unsound assumptions and argue unsoundly from them. But Melissus offends the more grossly of the two, so as really to raise no valuable point for discussion.

2. The false reasoning of Melissus is palpable; for, assuming that 'all that comes into existence has a beginning,' he deduces from it 'all that does not come into existence has no beginning.'

3. And, moreover, the assumption itself 'whatever comes into existence has a beginning' is untenable, in so far as 'began some-when' is taken (as Melissus takes it) to be equivalent to 'begins some-where' (so that if the Universe 'had no beginning' it 'can have no limit,' and is 'unbounded'): and again in so far as no distinction is made between the thing itself having to begin-to-be at 'some particular point of time,' and a modification of the thing having to start from 'some particular point within the thing itself'—as if there should not be a simultaneous modification over the whole field affected.

4. And again why should unity involve rigidity? For if a definite body of water, regarded as a unit without internal distinctions of quality, may have currents of motion within itself, why not the Universe? And, in any case, why not modifications other than those of local movement?

5. But of course the Universe cannot really be homogeneous, like a mass of water (except in the sense of its ultimate constituent being uniform; in which sense, though not in the other, some of the Physicists also have maintained its unity); for

---

[a]Includes 11. 186 a 4—186 a 21, Wicksteed and Cornford, *op. cit.*, pp. 26–31.

a man is patently different in kind from a horse, and opposites are different in kind from each other.

The argument of Melissus is resolved.

29(1). After the Philosopher has disproved the position of Parmenides and Melissus, *he begins to resolve their arguments*. What he does is threefold. First, he shows how their arguments are to be resolved; secondly, he resolves the argument of Melissus (2); thirdly, he resolves that of Parmenides (less. 6, n. 1).

30(2). First, he says that it is not difficult, therefore, to resolve the arguments from which Parmenides and Melissus syllogize, because both syllogize sophistically both in that they assume false proportions and in that they do not keep the proper form of the syllogism. But the argument of Melissus is *more irksome*, i.e., more vain and fatuous, and *it is not defective*,[52] i.e., it does not raise doubt. For he assumes what is contrary to natural principles and what is obviously false, viz., that being is not generated. Hence, it is not serious if, when one absurdity is presented, others follow.

31(3). Then, when he says: (2), he resolves the argument of Melissus, which was as follows: What is made has a beginning. What is not made does not, therefore, have a beginning.[a] But being is not made. It does not, therefore, have

[a]Cf. Melissus, frag. 1: "That which was, was always and always will be. For if it had come into being, it necessarily follows that before it came into being, Nothing existed. If however Nothing existed, in no way could anything come into

a beginning, and consequently, it does not have an end. But what does not have a beginning and an end is infinite. Being, therefore, is infinite. But what is infinite is immobile, for it does not have anything outside itself by which it might be moved. Again, what is infinite is one; because if it were many, there would have to be something outside the infinite. Being, therefore, is one, infinite and immobile.[a] But to show that being is not generated, Melissus brought in an

being out of nothing." Freeman, *op. cit.*, p. 48; frag. 2: "Since therefore it did not come into being, it Is and always was and always will be, and has no beginning or end, but it is eternal. For if it had come into being, it would have a beginning (for it would have come into being at some time, and so begun), and an end (for since it had come into being, it would have ended). But since it has neither begun nor ended, it always was and always will be and has no beginning nor end. For it is impossible for anything to Be, unless it Is completely." *Ibid.* The parentheses here are from the original text and not by Freeman.

[a]Cf. Melissus, frag. 7: "(1) Thus therefore it is everlasting and like throughout (homogeneous). (2) And neither could it perish or become larger or change its (inner) arrangement, nor does it feel pain or grief. For if it suffered any of these things, it would no longer be One. For if Being alters, it follows that it is not the same, but that that which previously Was is destroyed, and that Not-Being has come into being. Hence if it were to become different by a single hair in ten thousand years, so it must be utterly destroyed in the whole of time. (3) But it is not possible for it to be rearranged either, for the previous arrangement is not destroyed, nor does a non-existent arrangement come into being. And since it is neither increased by any addition, nor destroyed, nor changed, how could it have undergone a rearrangement of what exists? For if it were different in any respect, then there would be at once a rearrangement. (4) Nor does it feel pain; for it could not Be completely if it were in pain; for a thing which is in pain could not always Be. Nor has it equal power with what is healthy. Nor would it be the same if it were in pain; for it would feel pain through the subtraction or addition of something, and could no longer be the same. (5) Nor could that which is healthy feel pain, for the Healthy—That which Is—would perish, and That which Is Not would come into being. (6) And with regard to grief, the same reasoning applies as to pain. (7) Nor is there any Emptiness; for the Empty is Nothing; and so that

argument which some natural philosophers used. And so, Aristotle presents this argument below, near the end of the first book.[a]

32(4). He argues against this reasoning on four counts. First, he argues against the point: *What is made has a beginning. What is not made, therefore, does not have a beginning.* This, indeed, does not follow but is a fallacy of the consequent. For he argued from the destruction of the antecedent to the destruction of the consequent, while the correct form of argumentation is to argue conversely. And so, this does not follow: If it is made, it has a beginning; if it is not made, therefore, it does not have a beginning. What should follow is: If, therefore, it does not have a beginning, it is not made.

33(5). Secondly, at: (3), Aristotle disproves the foregoing argument with regard to this inference: *It does not have a beginning: it is, therefore, infinite.* Beginning, indeed, is said in two ways. In one way, it is said as a beginning *of time and of generation*. And it is taken in this sense, when this is said: What is

---

which is Nothing cannot Be. Nor does it move; for it cannot withdraw in any direction, but (all) is full. For if there were any Empty, it would have withdrawn into the Empty; but as the Empty does not exist, there is nowhere for it (Being) to withdraw. (8) And there can be no Dense and Rare. For the Rare cannot possibly be as full as the Dense, but the Rare must at once become more empty than the Dense. (9) The following distinction must be made between the full and the Not-Full; if a thing has room for or admits something, it is not full; if it neither has room for nor admits anything, it is full. (10) It (Being) must necessarily be full, therefore, if there is no Empty. If therefore it is full, it does not move." *Ibid.*, pp. 48–49.

[a]Less. 14, n. 2.

made has a beginning or what is not made does not have a beginning. In another way, it is said as a beginning *of the thing*[a] or of extension,[b] and thus this would follow: If it does not have a beginning, it is infinite. And so, it is clear that Melissus takes the word beginning as if it were said only in one way. And Aristotle says that it is absurd to say that the beginning of *everything*, i.e., of whatever has a beginning, is the *beginning of the thing*, i.e., of its extension, and not to say beginning in another way as the beginning of time and of generation.[c,53] Yet, he does not mean that simple and instantaneous generation that is the infusion of form into matter has a beginning, because there is no beginning of simple generation, but there is a beginning of the whole alteration whose terminus is generation, since it is not an instantaneous change, and sometimes this is called generation because of its terminus.

34(6). In the third place, at: (4), Aristotle disproves the above argument with regard to the third inference, where it is inferred: *It is infinite. It is, therefore, immobile.* And he shows that it does not follow in the case of local motion, because some part of water can be moved in the water itself[d] in such a way that it would not be moved to an extrinsic place, but it would be moved in respect to the congregation and dispersal of the parts.[54] And similarly, if a whole infinite body were water, it would be possible that its parts be moved within

---

[a]"rei."

[b]"magnitudinis."

[c]Cf. St. Thomas, *In V Metaph.*, less. 1, nn. 750–760.

[d]"seipsa."

the whole and not proceed outside the place of the whole. Likewise, he disproved it[a] in the case of the motion of alteration, because nothing would prevent the infinite from being altered either in its totality or in its parts, for this is not the reason why we have to posit something outside the infinite.

35(7). Fourthly, in: (5), he disproves the foregoing argument with regard to the fourth inference, where it was concluded *that if being is infinite, it is one.*[b]

---

[a]The above argument.

[b]Cf. Melissus, frag. 8: "(1) This argument is the greatest proof that it (Being) is One only; but there are also the following proofs: (2) If Things were Many, they would have to be of the same kind as I say the One is. For if there is earth and water and air and fire and iron and gold, and that which is living and that which is dead and black and white and all the rest of the things which men say are real: if these things exist, and we see and hear correctly, each thing must be of such a kind as it seemed to us to be in the first place, and it cannot change or become different, but each thing must always be what it is. But now, we say we see and hear and understand correctly, (3) and it seems to us that the hot becomes cold and the cold hot, and the hard soft and the soft hard, and that the living thing dies and comes into being from what is not living, and that all things change, and that what was and what now is are not at all the same, but iron which is hard is worn away by contact with the finger, and gold and stone and whatever seems to be entirely strong (is worn away); and that from water, earth and stone come into being. So that it comes about that we neither see nor know existing things. (4) So these statements are not consistent with one another. For although we say that there are many things, everlasting(?), having forms and strength, it seems to us that they all alter and change from what is seen on each occasion. (5) It is clear therefore that we have not been seeing correctly, and that these things do not correctly seem to us to be Many; for they would not change if they were real, but each would Be as it seemed to be. For nothing is stronger than that which is real. (6) And if it be changed, Being would have been destroyed, and Not-Being would have come into being. Thus, therefore, if Things are Many, they must be such as the One is." Freeman, *op. cit.*, pp. 49–50; frag. 9: "If therefore Being Is, it must

For it did not follow that it is one in species, but, perchance, in matter, as some natural philosophers stated that all things were one in matter but not in species. It is obvious, indeed, that man and horse differ in species; and similarly, contraries are mutually different in species.

---

be One; and if it is One, it is bound not to have body. But if it had Bulk, it would have parts, and would no longer Be." *Ibid.*, p. 50; frag. 10: "If Being is divided, it moves; and if it moved, it could not Be." *Ibid.*

1. The same kind of argument will apply to Parmenides, as well as such other arguments as particularly apply to his treatment; for here too the refutation turns on the falsity of his assumption and the unsoundness of his deductions.

2. His assumption is false inasmuch as he treats 'being' as having only one meaning, whereas in reality it has several.

3. And his inferences are false, because, even if we accepted such a proposition as 'nothing that is not white exists,' and if 'white' had only one meaning, still the white things would be many and not one. Obviously not one in the sense of a homogeneous *continuum*. Nor in the sense of a conceptual identity, for there remains a conceptual distinction between the subject in which the whiteness is seated, and the qualification of 'being white,' and that distinction does not involve the separate existence of anything alongside of 'that which is white'; because the plurality is established not by there being something separate, but by there being a conceptual distinction between white and the subject in which it inheres. But Parmenides had not yet arrived at this principle.

4. Parmenides, then, must assume not only that the word 'is,' whatever it may be predicated of, has only one meaning, but also that it means 'is identical-with-Being,' and that 'is one' means 'is identical-with Unity.' ('Being' will then no longer be regarded as an attribute); for an attribute is ascribed to some subject (other than itself); consequently, the subject to which 'being' (supposing it to be an attribute) is ascribed will have no being at all; for it will be other than 'being' (the attribute ascribed to it) and so will be something which (simply) is not. Accordingly 'identical-with-Being' (the sole meaning we have given to the word 'is') cannot be an attribute of some subject other than itself. For in that case its subject cannot be a thing which is, unless 'being' denotes a plurality of things in the sense

[a]Includes 11. 186 a 22—186 b 36, Wicksteed and Cornford, *op. cit.*, pp. 30–36.

that each is some thing that is. But it has been assumed that 'being' denotes only one thing.

5. If, then, what is identical-with-Being is not an attribute of anything else, but (a subject, so that other things are attributes) of it, is there any reason to say that 'identical-with-Being' denotes that which *is* (a real entity) any more than that which *is not* (a nonentity)? (I can show that there is no reason.) For suppose the thing which is identical-with-Being also has the attribute 'white,' and that 'being white' is not identical-with-Being—(the only sense in which it can 'be' at all), for it cannot even have 'being' as an attribute, because (*ex hypothesi*) nothing except what is identical-with-Being has any being at all—then it follows that white *is not*, and that not merely in the sense that it *is not this or that*, but in the sense of an absolute nonentity. Accordingly, that which is identical-with-Being (our subject) will be a nonentity; for (we assumed that) it is true to say of it that it is white, and this means that it is a nonentity. So that even if (to escape this difficulty) we say that the term 'white' denotes what is identical-with-Being, then 'being' has more than one meaning.

6. (Nor can 'the existent' have any magnitude, if it is to exclude plurality; for one part of it will have an existence distinguishable from that of another.)

7. But the analysing of a substantive entity into other substantive entities (so far from being anything startling) is clearly illustrated by definitions. For instance, if 'man' signifies a substantive existence, so must 'animal' and 'biped' also. Otherwise they would be accidental attributes, whether of 'man' or of some other subject. But that is impossible, for an accidental attribute must either be separable, so as sometimes to apply and sometimes not (*e.g.* that the man is 'sitting down'), or else must include its subject in its own definition (as 'snub' includes in its definition the definition of 'nose,' of which we say that snubness is an attribute). Now, the terms of the definition (which are the constituent principles of the thing defined) do not themselves severally contain, in *their* definitions, any mention of the whole thing which they combine to define. For instance, 'man' does not enter into the definition of 'biped,' nor does that of 'white man' enter into the definition of 'white.' Therefore, if 'biped' were an attribute of man at all it would have to be a separable one, so that the man might on occasion, not be a biped; the only alternative (as we have said) being that 'man' should be included in the definition of 'biped'; and this is not so, for it is the other way about, 'biped' being included in the definition of 'man.' And if both 'biped' and 'animal' were attributes of some

other subject than man and were not themselves, severally, subjects at all, then man himself would belong to the class of 'things attributed to a subject.' We must then, absolutely lay it down, that the substantively existent is not an attribute of something else; and also that what is true in this respect of the elements of a definition, severally and collectively, is also true of the thing which they define. The universe then is composed of a plurality of distinct individual entities.

The argument of Parmenides is resolved in a variety of ways.

36(1). After the Philosopher has argued against the reasoning of Melissus, *he argues against that of Parmenides*. First, he argues against it, and secondly, he excludes the remarks of those who wrongly answered the reasoning of Parmenides (less. 7, n. 1). With regard to the first, he proceeds in a twofold manner. First, he presents the methods by which the reasoning of Parmenides should be answered, and secondly, he resolves it by those methods (2).

37(2). With regard to the first, viz., the methods of answering the reasoning of Parmenides, it should be known that the argument of Parmenides was as follows (as is evident in *Metaphysics* I):[a,55] *Whatever is other than being is non-being; but what is non-being is nothing; therefore, whatever is other than being is nothing. But being is one; therefore, whatever is other than one is nothing; therefore, there is only one being.*[56] And from this he concluded that it would be immobile, because it would not have anything by which it might be moved, nor would it have anything outside itself to which it might move.[b]

[a]St. Thomas, less. 9, nn. 138–139. Cf. also Bk. V, less. 9, nn. 889 ff.

[b]"movetur."

From these arguments, moreover, it is evident that Parmenides considered *being* precisely from the viewpoint *of being*;[a,57] and for that reason he stated being to be one and finite. Melissus, on the other hand, considered being from the viewpoint of matter; for he considered being as what is made or not made, and for that reason, he stated being to be one and infinite.

38(3). Aristotle says, therefore, that the method of procedure against the argument both of Parmenides[b] and of Melissus is the same. For just as the

---

[a]"secundum rationem entis." We have italicized "being" and "of being."

[b]Cf. Parmenides' frag. 7–8: "For this (view) can never predominate, that That Which Is Not exists. You must debar your thought from this way of search, nor let ordinary experience in its variety force you along this way, (namely, that of allowing) the eye, sightless as it is, and the ear, full of sound, and the tongue, to rule; but (you must) judge by means of the Reason (Logos) the much-contested proof which is expounded by me.

There is only one other description of the way remaining, (namely), that (What Is) Is. To this way there are very many sign-posts; that Being has no coming-into-being and no destruction, for it is whole of limb, without motion, and without end. And it never Was, nor Will Be, because it Is now, and Whole all together, One, continuous; for what creation of it will you look for. How, whence (could it have) sprung? Nor shall I allow you to speak or think of it as springing from Non-Being; for it is neither expressible nor thinkable that What-Is-Not Is. Also, what necessity impelled it, if it did spring from Nothing, to be produced later or earlier? Thus it must Be absolutely, or not at all. Nor will the force of credibility ever admit that anything should come into being, beside Being itself, out of Not-Being. So far as that is concerned, Justice has never released (Being) in its fetters and set it free either to come into being or to perish, but holds it fast. The decision on these matters depends on the following: IT IS, or IT IS NOT. It is therefore decided—as is inevitable—(that one must) ignore the one way as unthinkable and inexpressible ((for it is no true way)) and take the other way as the way of Being and Reality. How could Being perish? How could it come into being? If it came into being, it Is Not; and so too if it is about-to-be at some future

argument of Melissus is resolved by pointing out the fact that he assumed false propositions, and the fact that he did not conclude properly in correct syllogistic form, so, too, the argument of Parmenides is resolved, partly because he assumes false propositions and partly because he does not conclude properly. Aristotle says, moreover, that there are other proper ways of refuting Parmenides, because we can refute him on the basis of propositions that he himself held, which are in some sense true and probable. But Melissus proceeded from something false and improbable, viz., that being is not generated. And this is why Aristotle did not employ propositions held by Melissus in arguing against him.

39(4). Then, when he says: (2), he pursues the foregoing methods. First, he employs the first one, and then the second: (3). First, he says, therefore, that Parmenides *assumes false propositions* because he takes *what is*, i.e., being, to be

---

time. Thus Coming-into-Being is quenched, and Destruction also into the unseen.

Nor is Being divisible, since it is all alike. Nor is there anything (here or) there which could prevent it form holding together, nor any lesser thing, but all is full of Being. Therefore it is altogether continuous; for Being is close to Being.

But it is motionless in the limits of mighty bonds, without beginning, without cease, since Becoming and Destruction have been driven very far away, and true conviction has rejected them. And remaining the same in the same place, it rests by itself and thus remains there fixed; for powerful Necessity holds it in the bonds of a Limit, which constraints it round about, because it is decreed by divine law that Being shall not be without boundary. For it is not lacking; but if it were (spatially infinite), it would be lacking everything.

To think is the same as the thought that It Is; for you will not find thinking without Being, in (regard to) which there is an expression. For nothing else either is or shall be except Being, since Fate has tied it down to be a whole and motionless; therefore all things that mortals have established, believing in their truth, are just a name; Becoming and Perishing, Being and Not Being, and Change of position and alteration of bright colour...." Freeman, *op. cit.*, pp. 43–44. Double parentheses contains what was in parenthesis in the original.

said *absolutely*, i.e., in one way, even though it may be said in many ways.[58] For in one way being is said to be substance and in another way accident, and it is said in a variety of ways according to the different genera. Being can also be taken as common to substance and to accident. It is evident that propositions that Parmenides held are in one sense true and in another sense false; for when it is said that *whatever is other than being is non-being*, it is true, if *being* is taken as common to substance and to accident. If, however, it is taken as accident only or as substance only, it is false, as will be shown below.[a] And, similarly, when he says that *being is one*, it is true, if it is taken as some one substance or as some one accident, but it will not be true, if taken to mean that whatever is other than that being is non-being.

40(5). Then, when he says: (3), he pursues the second method of solution, viz., that the reasoning of Parmenides *did not conclude correctly*. And first, he shows it by a comparison; secondly, he adapts it to the point at issue (4). First, he says, therefore, that from the fact that the form of argumentation of Parmenides is not cogent in every case, which would be necessary if it were a proper form of argumentation, it can be known that his reasoning does not conclude properly. For if we take *white* in place of *being*, and say that white signifies one only and is not said equivocally, and if we say that *whatever is other than white is not white, and whatever is not white is nothing*, it would not follow that white is one[59] only, first, indeed, because it will not be necessary that all white things be *one continuum*. Or, in another way, it will not be one white thing *by continuation*, i.e., from this very

---

[a]Nn. 7–9.

fact that it is something continuous, it will not be one simply, since the continuum is in a certain way many,[60] as was said above.[a]

And, similarly, it will not be one *in meaning*, for the meaning of white is different from the meaning of that of which white can be said. And yet, it will not be something other than white as if separated from it, for white is not something other than that of which white can be said for the reason that white is separable from that of which white can be said, but because the meaning of white is different from the meaning of that of which white can be said. But at the time of Parmenides it was not yet considered that something was one in subject and many in meaning.[b] And, therefore, he believed that if there is nothing outside a subject, it should follow that there is one.[c,61] But this is false, not only on account of the multitude of the parts, but also on account of the different meanings of subject and accident.[62]

41(6). Then, when he says: (4), he adapts the similitude to the point in question, i.e., what was said about white, and he shows that this pertains to being in a similar way. And with regard to this what he does is twofold. First, he shows that being is not one absolutely, because subject and accident are different in meaning; secondly, he shows the same thing based on the multitude of parts (6).

---

[a]Less. 3, middle of n. 3.

[b]For Parmenides even the intellect was made part of the physical *one*. Cf. frag. 16: "For according to the mixture of much-wandering limbs which each man has, so is the mind which is associated with mankind: for it is the same thing which thinks, namely the constitution of the limbs in men, all and individually; for it is excess which makes Thought." Freeman, *op. cit.*, p. 46.

[c]Cf. St. Thomas, *In V. Metaph.*, less. 7, n. 865.

With regard to the first, viz., showing that being is not one absolutely, he proceeds again in a twofold manner. First, he shows that when it is said that *whatever is other than being is non-being*, *being* cannot be taken as accident only; secondly, he shows that it cannot be taken as substance only (5).

42(7). First, he says, therefore, that when it is said that *whatever is other than being is non-being*, if *being* is said to have one meaning, it will be necessary that it signify not any being whatsoever, or be predicated of any being *whatsoever*, but it will be necessary that it signify *what truly is*, i.e., substance,[63] and that it signify *what truly is one*, viz., the indivisible. For, if being should signify accident, since accident may be predicated of a subject, it is necessary that the subject to which the accident called being occurred, would not be. For, if whatever is other than being is non-being, i.e., other than accident, and if the subject is different from the accident which signifies what I call being, it should follow that the subject is non-being.[64] And so, since an accident, which is being, would be predicated of a subject which is non-being, it should follow that being is predicated of non-being. And this is what Aristotle concludes. *And so something will exist when it does not exist*, as if he were to say: It should follow, therefore, that non-being is being. This, however, is impossible, because the first thing supposed in sciences is that contradictories may not be predicated of each other, as is said in *Metaphysics* IV.[a,65] And so, he concludes that if something is truly being, granted

[a]Less. 15, nn. 718–719; *ibid.*, less. 16, nn. 720–738; *ibid.*, less. 17, nn. 739–743. On the subject of first principles, cf. also St. Thomas, *In I Post. Anal.*, less. 5, n. 7; *ibid.*, less. 19–20; *In II Post. Anal.*, less. 20, nn. 1–2; *ibid.*, less. 40; *Summa Theol.*, Ia IIae, qu. 94, art. 2; *ibid.*, qu. 51, art. 1; *ibid.*, II IIae, qu. 1, art. 7; *In I Peri Hermeneias*, less. 15; *In IV Metaph.*, less. 6–8; *De Ver.*,

that *whatever is other that being is non-being*, then, that being is not an accident inhering in another. Because, then, it would not be true of the subject of this accident that it would in this way *be a being*, i.e., that the subject itself would not be strictly speaking being[a] unless being were to have many significations in such a way that any one of those many would be a being. But it is supposed by Parmenides that being has one signification only.

43(8). Then, when he says: (5), after he concluded that by *being* one cannot mean accident[b] when it is said that *whatever is other than being is non-being*, he shows that substance cannot be meant either. And so, he says, therefore, that if what truly is not an accident of something, but that something is an accident of it, in the proposition *whatever is other than being is non-being*, what should be signified by being rather than by non-being is *that which truly is*, i.e., substance.

But this cannot hold either, for let us assume that what is truly being, i.e., substance, is white. The white, however, is not what truly is, for it was already said[c] that what truly is cannot be an accident of anything. And this is said, therefore, because what is not truly, i.e., what is not substance, is not what is, i.e., it is not being. But whatever is other than being, i.e., other than substance, is non-being. It follows, therefore, that white is non-being. And not only is it non-being in such a way that it is not *this being*, as man is not this being which is an ass, but

---

qu. 11, art. 1; *ibid.*, qu. 8, art. 11; John of St. Thomas, *Curs. Theol.*, Tome VI, disp. 16, art. 1, beg. n. 11; Aristotle, *Peri Hermeneias*, chap. 9.

[a]"rationem entis."

[b]"intelligitur."

[c]N. 7, latter part.

that it does not exist at all, because Parmenides says that whatever is other than being is non-being, and what is non-being is nothing. From this, therefore, it follows that non-being may be predicated of what truly is, since white is predicated of substance which truly is, and yet, white does not signify being, as was said.[a] And so, it follows that being is non-being.[66] And this is impossible, because one contradictory may not be predicated of the other. And so, if, to avoid this absurdity, we say that true being signifies not only subject but also white itself, it follows that being would have many significations.[b,67] And in this way, it will not be one being only, since subject and accident are multiple in meaning.

44(9). Then, when he says: (6), he shows that on account of the multitude of parts it does not follow from the reasoning of Parmenides that there is one being only. And first, he shows this in relation to quantitative parts, and secondly, in relation to the parts of a definition (7).

First, he says, therefore, that if being were to have one meaning only, not only could it not be an accident with a subject, but neither will being be any magnitude, because every magnitude is divisible into parts, and each part does not have the same meaning. And so, it follows that the one being would not be a corporeal substance.[c,68]

---

[a]Above in this same par.

[b]Cf. W. Norris Clarke, S.J., "The Limitation of Act by Potency," *The New Scholasticism*, Vol. XXVI, No. 2 (April, 1952), 167–194.

[c]Cf. Parmenides, part of frag. 7–8: "But since there is a (spatial) Limit, it is complete on every side, like the mass of a well-rounded sphere, equally balanced from its centre in every direction; for it is not bound to be at all either greater or less in this direction or that; nor is there Not-Being which could check it from

45(10). Secondly, (7), he shows that being cannot be a *definable substance*, for it is obvious from the definition that what truly is, i.e., substance, has many divisions.[a,69] And any one of these is what truly is, i.e., substance. And these parts are different in meaning. In order that we may say that one thing which truly is is man, since man is a biped animal, animal must be and biped must be. Both of these will be what truly are, i.e., substances. But if they are not substances, they will, therefore, be accidents, either of man or of something else. But it is impossible that they be accidents of man.

And to show this he establishes two postulates. The first of these is that accident is said in two ways. In one way, an accident is separable, because it happens to be in the thing or not in the thing, e.g., sitting. In another way, an accident may be inseparable from and proper to a subject. And it is in the definition of this accident that is stated the subject of which it is an accident,[b,70] as, e.g., pugnosed is an accident peculiar to the nose, in the sense that nose must be stated in the definition of pug-nosed; for a pug nose is a curved nose.

The second postulate is that if some things are stated in the definition of something defined, or in the definition of any of the components of the definition,

---

reaching to the same point, nor is it possible for Being to be more in this direction, less in that, than Being, because it is an inviolate whole. For, in all directions equal to itself, it reaches its limits uniformly,..." Freeman, *op. cit.*, p. 44.

[a]Cf. Aristotle, *Categoriae*, chap. 5. The translations of this work that appear in the Appendix are those of E. M. Edghill in Richard McKeon's edition, *The Basic Works of Aristotle* (New York: Random House, 1941). Cf. also St. Thomas, *In III Metaph.*, less. 8, n. 433; *In V Metaph.*, less. 9, n. 889; *In XI Metaph.*, less. 1, n. 2169; *De Ente et Essentia*, chap. 1, ed. C. Boyer, S.J. (Rome: Gregorian University, 1933), pp. 10–12; *ibid.*, chap. 2, p. 13.

[b]*Ibid.*, chap. 7, pp. 53–55.

it is impossible that in the definition of any of these there be stated the definition of the defined whole, as, e.g., biped is stated in the definition of man, and as certain other things are stated in the definition of biped or animal from which man is defined. It is impossible, however, that man be stated in the definition of biped or in the definition of any of those things which are included[a] in the definition of biped or of animal.[71] Otherwise, it would be a circular definition, and the same thing would be both prior and posterior, and more known and less known, for every definition is from what is prior and more known, as is said in *Topics* VI.[b,72] And for this same reason, since white is stated in the definition of a white man, it is not possible that white man be stated in the definition of white.

These postulates having been made, he, therefore, argues as follows: If biped is an accident of man, either it must be a separate accident, and thus it would happen that man is not biped, which is impossible; or it will be inseparable, and thus man will have to be stated in the definition of biped, which is also impossible, because biped is stated in the definition of man. It is, therefore, impossible that biped be an accident of man, and for this same reason, neither is animal an accident of man.[c,73] But if it is said that both are accidents of something else, it would follow that man also would be an accident of something else. But this is impossible; for it was already said above[d] that what truly is is not an accident of

---

[a]"cadunt."

[b]Aristotle, chap. 14.

[c]Cf. St. Thomas, *De Ente et Essentia*, *op. cit.*, chap 3, pp. 18–21; *ibid.*, chap. 7, p. 53, par 2.

[d]Beginning of n. 8.

anything; but it is assumed that man is something which truly is,[74] as is clear from the above.[a]

Moreover, he makes it clear that it would follow that man is an accident of something else, if animal and biped are accidents of something else. He clarifies this in this way: Of whatever both animal and biped are said separately, both, i.e., *biped animal*, will be said together of that same thing. And of whatever we can say that it is *biped animal*, of that same thing[b] we can say that it is that which biped animal defined, viz., *man*, since man is nothing other than a biped animal.

So, therefore, if only one being is stated, it is evident that neither quantitative parts nor parts of magnitude nor parts of the definition can be stated.[75] Thus, it follows that every being is one indivisible, so that in stating one being, we would not be forced to say that it is many on account of its parts.

46(11). The Commentator says, however, that there (last two sentences of n. 7), Aristotle states a second argument of Parmenides to show that being is one. It is as follows: Being, which is one, is a substance and not an accident (and by substance he means body); if, however, that body is divided into two halves, it follows that being may be said of each half and of the composite of these. And this either proceeds to infinity, which is impossible, as he himself admits, or it will divide up to a certain point, which is also impossible. And so, being must be an indivisible one. But this explanation is distorted and contrary to the intention of Aristotle, as is apparent enough to anyone looking at the original text.[c]

---

[a]Middle of n. 10.

[b]"quod est ex eis."

[c]"litteram inspicienti secundum primam expositionem."

TEXT OF ARISTOTLE

LESSON SEVEN[a]

1. Note that some thinkers have given in to both the Eleatic arguments—to the argument that, if 'being' has only one meaning, all things are one, by conceding that 'what is not' exists; and also to the argument from dichotomy by supposing the existence of indivisible magnitudes.

2. Now it is obvious that, from the premises 'being has only one meaning' and 'contradictories cannot co-exist,' it is not a true inference that there is nothing which 'is not'; for 'what is not' may very well (not 'exist' absolutely, but) be 'what is *not this or that*.'

3. But to assert that, if there is to be nothing over and above 'just what is,' it will follow that all things are one, is absurd. For who would take this expression 'just what is' to mean anything but 'something that substantively exists'? But if it means that, there is nothing against the things that exist being a plurality, as we have seen.

It is clear, then, that the existent cannot be all 'one' in this sense.

---

[a]Includes 11. 187 a 1—187 a 11, Wicksteed and Cornford, *op. cit.*, pp. 36–38.

Those who said that non-being is something are proved wrong.

47(1). After the Philosopher has argued against the reasoning of Parmenides by showing the absurdities involved, *here he argues against the position of some who conceded the absurdities mentioned above*. And what he does is two-fold. First, he presents their position, and secondly, he argues against it: (2).

48(2). We must first consider, therefore, that above[a] the Philosopher employed two arguments against the reasoning of Parmenides. One was to show that from the argument of Parmenides it does not follow that all things are one on this account, viz., that subject and accident are different. And this first argument of Aristotle brought out this absurdity, viz., that non-being is being, an absurdity made manifest from the above.[b] But the other argument[c] proceeded to show that it does not follow that all things are one, for the reason, viz., that if it were a magnitude, it would follow that magnitude is indivisible, because if it were divisible, it would somehow be many.

---

[a]Less. 6, nn. 5 ff.

[b]*Ibid.*, nn. 6–9.

[c]*Ibid.*, n. 9–end.

49(3). The followers of Plato admitted both these arguments while granting the possibilities to which they led. They, therefore, admitted the first argument which involved saying that non-being would be being if one were to say that being has one meaning, either that of substance alone or accident alone and in so saying would mean that all things are one. I say that they admitted this argument, viz., that non-being would be being, for Plato said an accident is non-being. And for this reason, it is said in *Metaphysics* VI[a,76] that Plato stated that Sophistics was concerned with non-being, because this science deals especially with those things which are predicated accidentally. In this way, therefore, Plato, understanding substance by being, conceded the first proposition of Parmenides, who said that *whatever is other than being is non-being*, by stating that accident, which is other than substance, was non-being. Yet he did not concede the second proposition, viz., that *whatever is non-being is nothing*. For, although Plato said that an accident was non-being, still he did not say that an accident was nothing, but something. And on this account, according to him, it did not follow that being is one only.[b,77]

But Plato assented to the other argument which led to this, viz., that a magnitude was indivisible; he assented by making magnitudes *indivisible as a result of division*, i.e., by saying that the division of magnitudes terminates at

---

[a]Less. 2, nn. 1177 ff.; cf. also Plato, *Sophist*, 240–241, contained in Benjamin Jowett, *The Dialogues of Plato* (New York: Random House, 1920). All translations of dialogues of Plato that appear in the Appendix are taken from this work. Cf. also A. E. Taylor, *Plato* (New York: The Dial Press, Inc., 1936), pp. 389–390.

[b]Cf. Plato, 258, *ibid.*

indivisibles.[a,78] For Plato stated that bodies are resolved into surfaces and surfaces into lines, and lines into indivisibles, as is evident in *On Heaven and Earth* III.[b,79]

50(4). Then, when he says: (2), he argues against the foregoing position on this point, viz., that Plato conceded that non-being was something.[c,80] For in the proper place in the following sections[d] of natural philosophy, Aristotle argues against the fact that Plato said that there were individual magnitudes.[81] Moreover, he argues against the first point in two ways: first, he shows that it does not follow from the reasoning of Plato that non-being is something; secondly, he argues with regard to what Plato said, which is that if this is not stated (viz., that if non-being, which is an accident, is not something), it follows that all things are one (3).

51(5). He says first, therefore, that it is clear that it is not true that that argument of Plato from which he concluded that *being has one meaning* follows;[e] for Plato stated that being is a genus and is said univocally of all things according to their participation in the first being.[f,82] And again, he said that contradictories

---

[a]Cf. Plato, *Timaeus*, 53–54, *ibid.*; *Laws*, 893, *ibid.*; A. E. Taylor, *The Parmenides of Plato* (Oxford: Clarendon Press, 1934), pp. 37–39, 138–143; Francis M. Cornford, *Plato and Parmenides* (New York: Harcourt Brace and Co., 1939), pp. 198–199; St. Thomas, *In I Metaph.*, less. 16, n. 258.

[b]St. Thomas, less. 3–4.

[c]Cf. Plato, *Sophist*, 257, *op. cit.*

[d]Bk. VI, less. 1.

[e]It would not be a proper conclusion.

[f]Cf. Plato, *Sophist*, 247–248, *op. cit.*; Richard Robinson, *Plato's Earlier*

are not true simultaneously. From these two statements, he thought it followed th non-being was not nothing, but something; for, if being has one meaning, i.e substance, whatever is not substance will have to be non-being, because if it wer being, since being signifies substance only, it should follow that it is substanc And so, it would be substance and non-substance at the same time, i.e., contradi tories would be simultaneously true. If, therefore, it is impossible that contradi tories be true simultaneously, and if being has one meaning, viz., that of substanc it should follow that whatever is not substance is non-being. But there is somethin that is not substance, and this something is accident; therefore, there is somethin that is non-being. And so, it is not true that non-being is nothing.

Aristotle, however, shows that this does not follow, because, if being has on principal meaning, which is that of substance, nothing would prevent anyone fro saying that an accident, which is not substance, is not being in the full sense.[a] B this is not a reason why what is not something, i.e., substance, must be called fully non-being. And so, although accident is not being in the full sense,[c] still it cann be called fully[d] non-being.[83]

Dialectic (Oxford: Clarendon Press, 1953), pp. 257–264.

[a]"simpliciter."

[b]"absolute."

[c]"simpliciter." In this case, "simpliciter" may be a little stronger tha "absolute," but for all intents and purposes they are synonymous and, therefor may be translated by the same expression, such as "in the full sense." If o wished to show the distinction, one might translate it somewhat as follows: A so, although accident is not being in the absolute sense, it still cannot be calle unqualifiedly non-being.

[d]"absolute."

52(6). Then, when Aristotle says: (3), he shows further that it does not follow, if non-being, i.e., an accident, is not something, that all things are one. And this is what he means when he says it is illogical to say that all things are one, unless there is something outside being, because being can only mean substance, which is that which truly is.[a] But if there is substance, nothing can prevent it from being many, as was already said,[b] even if we eliminate magnitude[c] and accident, since the definition of substance is divided into many things which are in the genus of substance, as man is divided into animal and biped.[d,84] And it further follows that, according to the various differences of the genus, there are many actual[e] substances. And lastly, he draws the conclusion that he principally intended, viz., that all beings are not one, as Parmenides and Melissus said.

---

[a]"quae vere est."

[b]Less. 6, n. 10.

[c]Extension.

[d]And defined by Cf. *supra*, less. 6, n. 44.

[e]"in actu."

TEXT OF ARISTOTLE

LESSON EIGHT[a]

1. We turn now to the physicists. There are two schools of them. Those of the one school reduce existence to unity by positing a single underlying substance—whether one of the familiar three, or a something that is denser than fire and rarer than air—and arrive at a plurality by conceiving all else to be generated from it by condensation and rarefaction. Now dense and rare are opposites and may be brought under the more general conception of excess and defect. So Plato, too, has his 'great and small,' only he makes matter consist in this diversifying antithesis, and finds unity in the Idea; whereas the others find unity in the underlying matter and distinctions and forms in the opposites, or, like Anaximander, extract the contrasts themselves out of the indeterminate prime substance. The other school, to which Empedocles and Anaxagoras belong, start from the first with both unity and multiplicity; for they assume an undistinguished *confusum*, from which the constituents of things are sifted out. But they differ in this, that Empedocles supposes the course of Nature to return upon itself, coming round again periodically to its starting-point; while Anaxagoras makes it move on continuously without repeating itself. Moreover, he assumed an unlimited number of distinguishable substances, from the first, as well as an unlimited number of uniform particles in each substance; whereas Empedocles has only his four so-called elements.

[a]Includes 11. 187 a 12—187 a 31, Wicksteed and Cornford, *op. cit.*, pp. 41–42.

The opinions of the physicists who spoke in the manner of natural philosophers about principles.

53(1). After the Philosopher has argued against the opinion regarding the principles belonging to those who spoke about nature but not as a natural philosopher does, here *he takes up the opinions of those who spoke about the principles of nature from the point of view of natural philosophers, since they did not eliminate motion.* For that reason he calls them *physicists*, i.e., natural philosophers. With regard to this, what he does is twofold. First, he shows the difference in the opinions, and secondly, he takes up one of them (less. 9, n. 1).

54(2). First, he says, therefore, that in the opinion of natural philosophers, things may be generated from principles in two ways. One of these was employed by the natural philosophers who stated only one material principle, whether that was one of the three elements, viz., fire, air[a] and water (because nobody stated earth alone as a principle, as was said above),[b] or some medium between them, e.g., what was denser than fire and rarer than air. They said, however, that all other things were generated from this one principle on the basis of rarity and density; for

---

[a]"Anaximines of Miletus was in his prime about 546 B.C....One sentence only has survived." Frag. 2: "As our soul, being air, holds us together, so do breath and air surround the whole universe." Freeman, *op. cit.*, p. 19.

[b]Less. 2, n. 2, first part.

example, those who stated air as a principle, said that when this was rarefied, fire was generated from it, and that, moreover, when this was condensed, water was generated from it. Rare and dense, however, are contraries, and they are reduced to excess and defect, as to some more universal things, for the dense is what has much matter, while the rare is what has little matter.

55(3). And so, in a way they agreed with Plato, who said that the *great* and *small* were principles, for these also belong to excess and defect. But they differed from Plato in this, viz., that Plato put the great and small on the side of matte, because he stated one formal principle, which is a certain idea shared by things that are different in matter.[a,85] On the other hand, the ancient natural philosophers stated contrariety on the side of form, because they said that the first principle was one matter from which many things are constituted in virtue of different forms.

56(4). Other ancient natural philosophers, however, said that things were made out of principles, because contrary and different things were themselves drawn from one being in which they were as mixed together and confused. But they differed in that Anaximander[b] stated that the confused one, but not those things

---

[a]Cf. Plato, *Statesman*, 283-285, *op. cit.*; *Parmenides*, 131, 158, *ibid.*; *Philebus*, 24, *ibid.*; *Republic*, Bk. VII, 524, *ibid.*; *Phaedo*, 96, 104, *ibid.*; St. Thomas, *In VII Physic.*, less. 6, n. 939; *ibid.*, less. 8, n. 955; Léon Robin, *La théorie platonicienne des idées et des nombres d'après Aristote* (Paris: F. Alcan, 1908), pp. 635 ff.

[b]"Anaximander of Miletus was in his prime about 560 B.C. The title or titles of any works are unknown." Freeman, *op. cit.*, p. 19. Frag. 1: "The Non-Limited is the original material of existing things; further, the source from which existing things derive their existence is also that to which they return at their

mixed together in it, was a principle. And so, he stated only one principle. But Empedocles and Anaxagoras rather stated as principles what was mixed together in the confused one. And so, they stated many principles, although they, too, stated that the confused one was a principle in some sense.

57(5). But Anaxagoras and Empedocles differed in two ways. First, indeed, they differed because Empedocles stated a certain cycle of mixing together and of segregation. For he said that the world was often made and often destroyed, viz., in such a way that when the world was destroyed, love mixed all things into one[a]

---

destruction, according to necessity; for they give justice and make reparation to one another for their injustice, according to the arrangement of Time." Frag. 2: "This (essential nature, whatever it is, of the Non-Limited) is everlasting and ageless." Frag. 3: "(The Non-Limited) is immortal and indestructible." *Ibid.*

[a]Cf. Empedocles, frag. 16: "(Love and Hate): As they were formerly, so also will they be, and never, I think, shall infinite Time be emptied of these two." Frag. 17: "I shall tell of a double (Process): at one time it increased so as to be a single One out of Many; at another time again it grew apart so as to be Many out of One. There is a double creation of mortals and a double decline: the union of all things causes the birth and destruction of the one (race of mortals), the other is reared as the elements grow apart, and then flies asunder. And these (elements) never cease their continuous exchange, sometimes uniting under the influence of Love, so that all become One, at other times again each moving apart through the hostile force of Hate. Thus in so far as they have the power to grow into One out of Many, and again, when the One grows apart and Many are formed, in this sense they come into being and have no stable life; but in so far as they never cease their continuous exchange, in this sense they remain always unmoved (unaltered) as they follow the cyclic process.

But come, listen to my discourse! For be assured, learning will increase your understanding. As I said before, revealing the aims of my discourse, I shall tell you of a double process. At one time it increased so as to be a single One out of Many; at another time it grew apart so as to be Many out of One—Fire and Water and Earth and the boundless height of Air, and also execrable Hate apart from

and that when the world was again generated, strife separated and distinguished them; and so, distinction followed confusion and vice versa. But Anaxagoras said that the world was made only once, in such a way that from the beginning, all things were mixed together into one,[a] but that the intellect, which began to extract and distinguish, will never cease doing this, in such a way that all things will never

---

these, of equal weight in all directions, and Love in their midst, their equal in length and breadth. Observe her with your mind, and do not sit with wondering eyes! She it is who is believed to be implanted in mortal limbs also; through her they think friendly thoughts and perform harmonious actions, calling her Joy and Aphrodite. No mortal man has perceived her as she moves in and out among them. But *you* must listen to the undeceitful progress of my argument.

All these (Elements) are equal and of the same age in their creation; but each presides over its own office, and each has its own character, and they prevail in turn in the course of Time. And besides these, nothing else comes into being, nor does anything cease. For if they had been perishing continuously, they would Be no more; and what could increase the Whole? And whence could it have come? In what direction could it perish, since nothing is empty of these things? No, but these things alone exist, and running through one another they become different things at different times, and are ever continuously the same." *Ibid.*, pp. 53–54.

[a]Anaxagoras, frag. 4: "Conditions being thus, one must believe that there are many things of all sorts in all composite products, and the seeds of all Things, which contain all kinds of shapes and colours and pleasant savours. And men too were fitted together, and all other creatures which have life. And the men possessed both inhabited cities and artificial works just like ourselves, and they had sun and moon and the rest, just as we have, and the earth produced for them many and diverse things, of which they collected the most useful, and now use them for their dwellings. This I say concerning Separation, that it must have taken place not only with us, but elsewhere.

Before these things were separated off, all things were together, nor was any colour distinguishable, for the mixing of all Things prevented this, (namely) the mixing of moist and dry and hot and cold and bright and dark, and there was a great quantity of earth in the mixture, and seeds infinite in number, not at all like one another. For none of the other things either is like any other. As this was so, one must believe that all Things were present in the Whole." *Ibid.*, p. 83.

be mixed into one. In another way, they differed in this, viz., that Anaxagoras stated an infinite number[a] of similar and contrary parts to be principles; for example, he stated an infinite number of parts of flesh that are like one another, and an infinite number of parts of bone and other things that have similar parts, even though there is contrariety of some things to others, just as there is contrariety between the parts of bone and the parts of blood based on the moist and the dry. But Empedocles stated that principles were only those four that are commonly called elements, viz., fire, air, water and earth.[b]

---

[a]Cf. Anaxagoras, frag. 7: "So that the number of the things separated off cannot be known either in thought or in fact." *Ibid.*, p. 84.

[b]Cf. Empedocles, frag. 38: "Come now, I will first tell you of (the sun) the beginning, (the Elements) from which all the things we now look upon came forth into view: Earth and the sea with many waves, and damp Air, and the Titan Aether which clasps the circle all round." *Ibid.*, p. 57; frag. 71: "But if your belief concerning these matters was at all lacking—how from the mixture of Water, Earth, Aether and Sun (Fire) there came into being the forms and colours of mortal things in such numbers as now exist fitted together by Aphrodite...." *Ibid.*, p. 59; frag. 98: "The Earth having been finally moored in the harbours of Love, joined with these in about equal proportions: with Hephaestus, with moisture, and with all-shining Aether, either a little more (of Earth) or a little less to their more. And from these came blood and the forms of other flesh." *Ibid.*, p. 62; frag. 107: "For from these (Elements) are all things fitted and fixed together, and by means of these do men think, and feel pleasure and sorrow." *Ibid.*, p. 63; frag. 109: "We see Earth by means of Earth, Water by means of Water, divine Air by means of Air, and destructive Fire by means of Fire; Affection by means of Affection, Hate by means of baneful hate." *Ibid.*

TEXT OF ARISTOTLE

LESSON NINE[a]

1. Anaxagoras appears to have based his conviction that the primal substances are unlimited in number on his uncompromising acceptance of the dogma, common to all the Physicists, that 'nothing can come out of what does not exist.' This made him declare that originally 'all things existed together' and explain that genesis was nothing more than the modification induced by setting them in order; whereas the same dogma made the others attribute genesis to transforming combination and resolution.

2. Further, Anaxagoras argued from the genesis of unlikes from each other that they were already in each other;

3a. for since whatever comes to be must arise either out of what exists or out of what does not exist, and since the latter was universally held to be impossible, it remained that all things arose out of what existed, and so must be there already, only in particles so minute as to escape our senses. So he and his school argued that particles of everything must exist in everything else, since they saw all kinds of things emerging from each other.

3b. And they held that things have a different appearance and receive different names according to the prevalence of one constituent or another in the mixture; and that accordingly, no such thing exists as pure black or white or sweet or flesh or bone, but the nature of a thing is judged by what it has most of in it.

4. But (1) if a thing has no limit under some certain aspect, then we cannot define it in respect of that said aspect as to which it is unlimited. Thus we cannot say how great a thing is, or how many of it there are, if it has no limit as to size or number; nor can we say what kind of thing it is if there is no limit to its diversity. If, then, the constituents of a thing are unlimited both as to magnitude and as to kind, we cannot know what that thing is; for we reckon to know about

---

[a]Includes 11. 187 a 33—188 a 18, Wicksteed and Cornford, *op. cit.*, pp. 43–51.

a thing that is put together when we know the quality and quantity of its components.

5. Again, (2) a thing, the parts of which can be of any magnitude, great or small, can itself be of any magnitude—I am speaking of parts into which, as already existing in it, the whole can be resolved. If this is a necessary consequence, and if it is impossible that an animal or plant should exceed all limit in greatness or smallness, the same must be true of any part of it; for the whole and the parts must bear a definite proportion to each other. Now flesh and bone and so forth are parts of animals, and fruits parts of plants. It is clear then that neither flesh nor bone nor anything of the kind can be great or small beyond limit.

6. (3) Further, if on the one hand (as Anaxagoras holds) all such things are already there in each other and do not come into existence but are merely sifted out from where they are and take their names from their dominant constituents, and anything can be sifted out of anything (water out of flesh or flesh out of water), but, on the other hand, any limited body must use up any other limited body in such a process, then clearly it is impossible for every thing to exist in every thing.

7. For if flesh has been sifted out of a given body of water, and then more flesh is sifted out of the remaining water, even if the successive extracts constantly diminish in quantity they cannot diminish below the minimum particle. Consequently, if, when they reach this minimum, the process is of necessity arrested, then it will no longer be true that everything is contained in everything, for there will be no flesh in the water that is left over. If, on the other hand, you can always go on sifting out one minimum particle at a time, it follows that there are an unlimited number of bodies of equal size contained in a limited body, which is impossible.

8. (4) Besides, since the subtraction of anything from a given body must reduce the size of that body, and since a mass of flesh cannot be indefinitely great or small, it is clear that from the minimum of flesh no other body can be extracted, for that would reduce it below its minimum.

9. (5) Again, in each of the unlimited number of primal substances an unlimited amount of flesh and blood and brains would exist, not indeed gathered together in recognizable aggregates, but still existing. So that each of these

substances would exist without limit within each of the others, which would also exist without limit within it. And this is impossible.

10. And yet this notion of something being there that can never become completely distinct is very sound, though Anaxagoras had not got the right hold of it. For modifications cannot exist apart, by themselves. So if such things as colours and states were included in the primal Anaxagorean *confusum* and could have been wholly disengaged from it, there would have resulted a 'white' and a 'healthy' that was nothing else but itself—not even to the extent of having any subject! So it was just as well that 'Intelligence' did not desire to sever such things out; for it could not be done, either with respect to quantity, since there is no minimum magnitude, nor with respect to quality, for modifications cannot exist apart by themselves.

11. Nor was his conception of genesis from similar particles adequate; for although in one way a piece of mud is divided into smaller pieces of the same stuff—mud, in another way it is not (but is divided into earth and water—dissimilar parts). Also, water and air do not consist of one another or come out of one another in the same way that bricks come out of a house that is broken up, or a house consists of bricks.

12. Empedocles, then, was, so far, sounder in assuming a small limited number of prime substances.

The opinion of Anaxagoras regarding infinite principles is attacked.

58(1). Now that he has stated the different opinions of natural philosophers[86] regarding principles, *here he studies one of these opinions, viz., that of Anaxagoras* because this opinion seemed to assign the common cause of all the kinds of motion. And this study is divided into two parts. In the first, he states the argument of Anaxagoras, and in the second, he brings arguments against it (4). He deals with the first in three ways. First, he premises what Anaxagoras postulated, and that from which he argued; secondly, he states the procedure of his reasoning (3); thirdly he gives his own response to a certain unmentioned objection (3b).

59(2). Now, Anaxagoras established two postulates and proceeded from these. The first of these was also postulated by all the natural philosophers, viz., that nothing comes to be from nothing. And this is what he means when he says that as a result Anaxagoras seemed to think that there were infinite principles, because he accepted as true the common opinion of all the natural philosophers, viz., that what absolutely is not, in no way comes to be. Indeed, since the natural philosophers postulated that as a principle, they proceeded to different opinions.

60(3). So that they would not be forced to say that anything new, which in no way existed before, came into being, some said that all things formerly had existed together, whether in some confused one, as Anaxagoras[a] and Empedocles[b] said, or in some material principle, as water, fire, and air, or in some medium between these. And accordingly, they stated two ways of coming into being. For some stated that all things pre-existed together as if in one material principle, and they said that coming into being was nothing other than being altered, for they said that all things came into being from one material principle by its condensation and rarefaction. But others, who said that all things pre-existed together, as if in some confused one and as if mixed together from many things, said that the coming into being of things is nothing more than congregation and segregation. And all these people were deceived, because they did not know how to distinguish between potency and act, for being in potency is as a medium between pure non-being and being in act. Those things that come into being naturally do not, therefore, come into being from what is non-being absolutely, but from being in potency, and not from being in act, as they thought. And so things which come into being do not have to pre-exist in act, as they said, but only in potency.

---

[a]Anaxagoras, frag. 17: "The Greeks have an incorrect belief on Coming into Being and Passing Away. No Thing comes into being or passes away, but it is mixed together or separated from existing Things. Thus they would be correct if they called coming into being 'mixing', and passing away 'separation-off.'" Freeman, *op. cit.*, p. 85.

[b]Empedocles, frag. 7: "(The Elements): uncreated." Frag. 8: "And I shall tell you another thing: there is no creation of substance in any one of mortal existences, nor any end in execrable death, but only mixing and exchange of what has been mixed, and the name 'substance' (Phusis, 'nature') is applied to them by mankind." *Ibid.*, p. 52.

61(4). Then, when he says: (2), he states the second point that he supposed. For he said that contraries come into being from one another, for we see that the cold comes into being from the warm, and vice versa.[87] And he concluded from this that, since nothing comes into being from nothing, this is true according to potency, for the cold is potentially in the hot, but not actually, as Anaxagoras thought,[a] because he did not know how to understand being in potency, which is being mediate between pure non-being and being in act.

62(5). Then, when he says: (3a), he states how he proceeds with his argument. And he proceeded in this way. If anything *comes into being*, it must come into being from being or from non-being. But because of the foregoing common opinion[b] of the philosophers, he excluded one of these, viz., the fact that something should come into being from non-being. And so, he concluded the remaining alternative, viz., that something should come into being from being, e.g., if air comes into being from water, that air existed previously. It would not be said, however, that air comes into being from water, unless air pre-existed in water. And so, he wished to hold that everything that comes into being from something, pre-existed in that from which it came into being.

But this seemed against what is apparent to the senses, for it is not sensibly apparent that what was generated from something pre-existed in it.[c] For that reason

---

[a]Anaxagoras, frag. 8: "The things in the one Cosmos are not separated off from one another with an axe, neither the Hot from the Cold, nor the Cold from the Hot." *Ibid.*, p. 84.

[b]N. 2.

[c]We might reflect upon the question of Anaxagoras in frag. 10: "How can hair come from not-hair, and flesh from not-flesh?" Freeman, *op. cit.*, p. 84. It

he excluded this objection by saying that what comes into being from something, pre-existed in it in certain minimal parts, which are imperceptible because of their smallness;[88] for example, if air comes into being from water, some minimal parts of air are in water, but not in that quantity in which it was generated. He said, therefore, that air came into being by the congregation of those parts of air and by their segregation from the parts of water.

Since he held this, viz., that everything that comes into being from something pre-existed in it, therefore, he assumed further that everything comes into being from everything. And so, he concluded that anything and everything was thoroughly mixed with anything and everything in virtue of minimal parts that are not perceptible. And because one thing can come into being from another an infinite number of times, he said that there were infinite minimal parts in any one thing.

63(6). Then, when he says: (3b), he excludes an unmentioned objection. For someone could object that if infinite parts of anything whatsoever are in anything whatsoever, it should follow that things neither differ from one another, nor would they appear to differ from one another. Thus, as if in answer to this, he says that things seem to differ from one another and are also called different by reason of their predominant constituent, even though there is an infinite multitude of minimal parts which are contained in any mixed thing. And so, nothing is purely and totally white or black or bone, but its predominant constituent seems to be the nature of the thing.[a,89]

---

certainly is not sensibly apparent that hair comes from what is not hair.

[a]Cf. Anaxagoras, frag. 11: "In everything there is a portion of everything

64(7). Then, when he says: (4), he argues against the foregoing position. And he deals with this in two ways. First, he argues against it absolutely, and secondly, he compares it to the opinion of Empedocles (12). With regard to the first, what he does is twofold. First he states the reasons for arguing against the opinion of Anaxagoras, and secondly, he argues against this manner of stating it (10). He states five arguments for the first. The first of these arguments follows: Everything infinite as infinite is unknown.[a] And he deliberately states why he says *as infinite*, because if it is infinite in multitude or magnitude, it will be unknown in quantity. If, however, it is infinite in species, e.g., what is constituted from infinite things that are different in species, then it will be unknown in quality. And the reason for this is that what is intellectually[b] known is intellectually comprehended with regard to all that belongs to it, and this cannot happen in the case of something infinite. If, therefore, the principals of anything are infinite, they must be unknown either in quantity or in species.

But if the principles are unknown, what is derived from the principles must be unknown.[91] He proves this from the fact that we think we know some one composite thing *when we know of what and of how much it is constituted*, i.e., when we know both the species and the quantities of the principles.[92] Thus, it follows from the first to the last, that if the principles of natural thing are infinite, natural things will be unknown, either in quantity or in species.[c]

---

except Mind; and some things contain Mind also." *Ibid.*

[a]Cf. *infra*, Bk. III, less. 6, 11, n. 383; cf. also St. Thomas, *Summa Theol.*, Ia, qu. 7, art. 1.

[b]"apud intellectum."

[c]Cf. St. Thomas, *In IV Sent.*, disp. 12, qu. 1, art. 1, sol. 3, ans. 1; *In VII*

65(8). He states the second argument (5). And it follows: If the parts of a whole do not have any determined quantity, be it large or small, but if they happen to have an indefinite size, either large or small, the whole must not have a determined size, be it large or small. But it so happens that the whole has an indefinite size, large or small. And this is so because the quantity of the whole comes from the parts.[a] (But this is to be understood with regard to parts which actually exist in the whole, as flesh, sinew, and bone in an animal. And this is what he means when he says, *but I say that a whole is divided into some one of each parts when that part is in the whole*, i.e., when it is actually in it. And by this he excludes the parts that are potentially in a continuous whole.[b,93] But it is impossible that an animal or plant or something of that kind be indeterminately disposed to an indefinite size, large or small, for there is a quantity so great beyond which no animal is extended, and a quantity so small below which no animal is found. And the same is true of plants. He proceeds, therefore, to the destruction of the consequent, viz., that no part has quantity indeterminately, because what is true of the whole is true of the parts.[94] But flesh and bone and such things are parts of the animal, and fruits are parts of plants. It is impossible, therefore, that flesh and bone and such things have an indeterminate quantity, either large or small. Thus, it is not possible that there be some parts of flesh or of bone which are insensible because of their smallness.[95]

Physic., beg. of less. 4.

[a]Cf. Anaxagoras, frag. 3: "For in Small there is no Least, but only a Lesser: for it is impossible that Being should Not-Be; and in Great there is always a Greater. And it is equal in number to the small, but each thing is to itself both great and small. Freeman, *op. cit.*, p. 83.

[b]Cf. St. Thomas, *In VII Metaph.*, less. 9, n. 960.

66(9). It seems, however, that what is said here is contrary to the division of a continuum to infinity. For, if a continuum is divisible to infinity, and flesh is a continuum, it seems that it would be divisible to infinity. In division to infinity, therefore, a part of flesh transcends every determined smallness. But we ought to say that although a body, taken mathematically, is divisible to infinity, yet a natural body is not divisible to infinity, for in mathematical body only quantity is considered. In this there is nothing repugnant to division to infinity, but in the natural body the natural form is considered which requires a determined quantity, just as other accidents do. Quantity, therefore, can be in the species flesh only if it is determined within certain limits.

67(10). He states his third argument (6). With regard to this what he does is twofold. First, he states some premises from which he argues, and secondly, he states the way he proceeds with his argument (7). With regard to the first, he proposes three points. The first is that all things are together, according to the position of Anaxagoras, as was said.[a] From this he wishes to argue to an absurdity. For Anaxagoras, as was mentioned,[b] said that all *such things*, viz., those that have similar parts, as flesh and bone and similar things, are in each other and do not come into being anew, but are separated out from something in which they pre-existed.[c] But any one thing is given its name *from its predominant*

---

[a]N. 3, first part.

[b]N. 3, second part.

[c]Cf. Anaxagoras, frag. 5: "These things being thus separated off, one must understand that all things are in no wise less or more (for it is not possible for them to be more than All), but all things are forever equal (in quantity)." Freeman, *op. cit.*, p. 83. The first parenthesis is from the original text.

constituent, i.e., from the majority of the parts existing in the thing. The second point is that anything whatsoever comes into being from anything whatsoever, just as water comes into being from flesh by segregation, and similarly, flesh from water. The third point is that every finite body *is cut off from a finite body*. This means that if there is repeatedly taken from some large finite body, no matter how large, a finite body, no matter how small, the less could be taken from the greater so many times that by its division the greater whole would be consumed by the lesser. From these three points, however, he concludes what he principally intends, viz., that each thing is not in each other thing. And this is contrary to the first of these three statements. In this way, indeed, in arguments leading to the impossible, the conclusion of arguments of this kind is that one of the premises is destroyed.

68(11). Then, when he says: (7), he derives an argument, and postulates what he had concluded in the preceding argument. He says, indeed, that if flesh is removed from water (i.e., while the flesh is being generated from water), and if again another segregation of flesh is made from the remaining water, although a lesser quantity of flesh always remains in the water, yet, the size of the flesh *does not become less than some smallness*,[a] i.e., there happens to be some small measure of flesh than which no flesh will be less, as is apparent from the above reasoning.[b,96] Now that it is established that there is some small amount of flesh than which there is none smaller, he proceeds as follows: If from water flesh is segregated, and again other flesh, that segregation will either stop or not. If it stops, there will be no flesh in the remaining water, and so, anything whatsoever

---

[a]"magnitudo carnis non *excedit aliquam parvitatem*."

[b]N. 8.

will not be in anything whatsoever.[97] If, however, it does not stop, then some part of flesh will always remain in the water, yet in such a way that in the second segregation it would be less than in the first, and in the third less than in the second. And since it is not possible to descend to infinity in smallness of parts, as was said,[a] these minimal parts of flesh will be equal and infinite in number in some finite water. Otherwise, segregation would not proceed infinitely. And so it follows, if segregation does not stop, and if flesh is always removed to infinity from water, that in some finite magnitude, viz., water, there are some things finite in quantity and equal to one another and infinite in number, viz., infinite minimal parts of flesh. And this is impossible and contrary to what was stated above,[b] viz., that every finite body must be cut off from some finite body. And, therefore, the first statement was impossible, viz., that anything whatsoever is in anything whatsoever, as Anaxagoras stated.[98]

69(12). Moreover, we must consider that it is not without reason that the Philosopher added *equal* in this last absurdity which he[c] brings out. For it is not absurd that in some finite being there are infinite things that are unequal, if the meaning of quantity is observed, because if a continuum is divided in the same proportion, it will be possible to proceed to infinity; for example, if we should take a third of a whole and a third of a third and so on, still, the parts taken will not be equal in quantity. But if we make a division into equal parts, it will not be possible

---

[a]Nn. 7–10.

[b]N. 10.

[c]Aristotle.

to proceed to infinity, even if we consider only the meaning of quantity in a mathematical body.

70(13). He states the fourth argument (8), which follows: Each body, when something is removed, becomes less, since every whole is greater than its parts. But since a quantity of flesh is determined in size, large and small, as is evident from what has been said,[a] there must be some minimal part of flesh. For this reason, something cannot be segregated from this because in this way something would then be less than the least. Anything whatsoever cannot, therefore, come into being from anything whatsoever by segregation.[99]

71(4). He states his fifth argument (9), which follows. If infinite parts of everything are in everything else, and if anything whatsoever is in anything whatsoever, it should follow that in infinite bodies there are infinite parts of flesh and infinite parts of blood or of brain, and no matter how much they are separated,[100] still they remain there. It should follow, therefore, that infinite bodies are infinitely in infinite bodies, and this is unreasonable.

72(15). Then, when he says: (10), he argues against the foregoing position of Anaxagoras with regard to the manner of stating it.[b,101] And he does this for two reasons. The first of these is that Anaxagoras did not understand his own position, and the second is that he did not have a sufficient motive for stating his

---

[a]N. 11.

[b]Cf. St. Thomas, *In I Metaph.*, less. 4, nn. 90–91; *ibid.*, less. 10, n. 152; *ibid.*, less. 12, nn. 194–196.

position (11). First he says, therefore, that when he stated that segregation might never be ended, he did not realize what he said, although in some sense he stated the truth,[a],[102] because accidents can never be separated from substances, and still, he stated the thorough mixture not only of bodies, but also of accidents.[103] For, when something becomes white, he said that this was done by the drawing out[b] of whiteness which was first mixed together. If, therefore, colors and other accidents of this kind may be said to be mixed together, as he himself said, and if anyone who granted this should say that all things that are mixed together can be segregated, it would follow that there would be a white and a healthy without any subject of which these are said, and in which they are, which is impossible. It remains, therefore, that it is true that not all things that are mixed together can be segregated, if accidents are also mixed together.

But an absurdity results from this, for Anaxagoras said that all things were mixed from the beginning, but that the intellect began to segregate.[c,104] But any

---

[a]*Ibid.*, nn. 194–199.

[b]"abstractione": taking away, "abstracting."

[c]Cf. Anaxagoras, frag. 14: "Mind, which ever Is, certainly still exists also where all other things are, (namely) in the multiple surrounding (mass) and in the things which were separated off before, and in the things already separated off." Freeman, *op. cit.*, p. 85; frag. 12: "Other things all contain a part of everything, but Mind is infinite and self-ruling, and is mixed with no Thing, but is alone by itself. If it were not by itself, but were mixed with anything else, it would have had a share of all Things, if it were mixed with anything; for in everything there is a portion of everything, as I have said before. And the things mixed (with Mind) would have prevented it, so that it could not rule over any Thing in the same way as it can being alone by itself. For it is the finest of all Things, and the purest, and has complete understanding of everything, and has the greatest power. All things which have life, both the greater and the less, are ruled by Mind. Mind took command of the universal revolution, so as to make (things) revolve at the outset.

intellect that seeks to do what cannot be done is an improper[a,105] intellect. That intellect that tends toward the impossible will, therefore be an absurdity, *if it truly wishes*, i.e., if it absolutely wishes to segregate. This, indeed, is impossible both in quantity, because there is no smallest magnitude, as Anaxagoras said,[b] (but from

---

And at first things began to revolve from some small point, but now the revolution extends over a greater area, and will spread even further. And the things which were mixed together, and separated off, and divided, were all understood by Mind. And whatever they were going to be, and whatever things were then in existence that are not now, and all things that now exist and whatever shall exist—all were arranged by Mind, as also the revolution now followed by the stars, the sun and moon, and the Air and Aether which were separated off. It was this revolution which caused the separation off. And dense separates from rare, and hot from cold, and bright from dark, and dry from wet. There are many portions of many things. And nothing is absolutely separated off or divided the one from the other except Mind. Mind is all alike, both the greater and the less. But nothing else is like anything else, but each individual thing is and was most obviously that of which it contains the most." *Ibid.*, pp. 84–85; frag. 13: "And when Mind began the motion, there was a separating-off from all that was being moved; and all that Mind set in motion was separated (internally); and as things were moving and separating off (internally), the revolution greatly increased this (internal) separation. *Ibid.*, p. 85.

It might prove interesting to compare the above fragments of Anaxagoras with frag. 110 of Empedocles: "...For all things, be assured, have intelligence and a portion of Thought." *Ibid.*, p. 64.

[a]"indecens," i.e., an intellect that does not function according to its nature. Cf. Plato, *Phaedo*, 97, *op cit.*

[b]Cf. Anaxagoras, frag. 6: "And since there are equal (quantitative) parts of Great and Small, so too similarly in everything there must be everything. It is not possible (for them) to exist apart, but all things contain a portion of everything. Since it is not possible for the Least to exist, it cannot be isolated, nor come into being by itself; but as it was in the beginning, so now, all things are together. In all things there are many things, and of the things separated off, there are equal numbers in (the categories) Great and Small." Freeman, *op. cit.*, p. 84.

he smallest thing whatsoever, something can be taken away), and also in quality, because accidents cannot be separated from their subjects.

73(16). Then, when he says: (11), he argues against the above proposition from the viewpoint that Anaxagoras did not have a sufficient motive, for, since Anaxagoras saw that something is made large from a congregation of many small similar parts, as a torrent becomes large from many drops of water, he believed this to be so in all things. And for this reason Aristotle says that Anaxagoras did not correctly understand the *generation of similar species*, i.e., that something always had to be generated from things similar in species,[106] for some things are generated from[107] things that are similar, and are resolved into similar things, as clay is divided into pieces of clay. In some cases, however, this is not so, for some things are generated from dissimilar things. And even among these there is not just one way, because some things come to be from dissimilar things by alteration, as bricks do not come to be from bricks, but from clay, and some things come to be by composition, as a house does not come to be from houses, but from bricks. And in this way air and water come to be from each other, i.e., as from things that are dissimilar. Another text states *as bricks from a house*. And so, he stated a double method by which something comes to be from things that are dissimilar, viz., by composition, as a house comes to be from bricks, and by resolution, as bricks come to be from a house.

74(17). Then, when he says: (12), he argues against the position of Anaxagoras in contrast to the opinion of Empedocles. And he said that it is better that principles be fewer and finite,[a,108] as Empedocles made them, rather than many and infinite as Anaxagoras made them.[b]

---

[a]Cf. Empedocles, frag. 39: "If the depths of the earth were unlimited, and also the vast Aether, a doctrine which has foolishly issued forth off the tongues of many, and has been spread abroad out of their mouths, since they have seen only a little of the Whole...." *Ibid.*, p. 57.

[b]Cf. Anaxagoras, frag. 2: "Air and Aether are separated off from the surrounding multiplicity, and that which surrounds is infinite in number." *Ibid.*, p. 83.

1. This brings us to observe that all these thinkers assume as principles some 'couple' of antithetical qualities or forces, and this whether they declare the sum of things to be one and rigid (for even Parmenides erects 'hot' and 'cold,' which he calls 'fire' and 'earth,' into principles); or whether they speak of 'rare' and 'dense'; or whether with Democritus, they speak of 'solidity' and 'vacancy' (the one regarded as the 'existent' and the other as the 'non-existent'), and further distinguish the atoms by position, shape, and order, all of which are expressed in antithetical couples: position as 'above and below,' 'before and behind,' and shape as 'angular, straight, or curved.'

Clearly, then, they all assume certain numbers of antithetical couples as principles;

2. and not without reason, for 'principles,' being themselves primary, must not be derived either from each other or from anything else, and all other things must arise out of them. And the terms of a primary antithesis fulfil this condition; for, because they are primary, they cannot be derived from anything else, and because they are antithetical, they cannot rise out of each other. But we must go more closely into the question what this means and how it works out.

Note, then, to begin with, that things cannot act upon each other and turn into one another at random, unless it be by incidental concomitance; for how (to take an example) could culture, as such, 'become' pallor? A man of culture, who had been of a swarthy complexion, might indeed become pale, and so the person we had described as 'cultured' might come to be a person we described as 'pale,' but only incidentally to the fact that the cultured person who became pale had other qualities concomitantly with his culture; for he cannot 'become' pale on the mere strength of already having some quality other than pallor to start from, unless that quality is on the specific line of antithesis to pallor—swarthiness or some intermediate shade. So too a man cannot 'become' cultured merely because he already has some characteristic which is not culture, unless it is the specific characteristic

---

[a]Includes 11. 188 a 19—189 a 10, Wicksteed and Cornford, *op. cit.*, pp. 51–59.

of complete or unpartial want of culture. And all this holds of the loss of qualifications just as much as of their acquisition. A man does not become cultured when he ceases to be pale (unless it be by incidental concomitance), for what 'ceasing to be pale' means is not only becoming something not pale, but something on the line antithetical to pallor that leads to swarthiness. So too, if culture lapses from itself, it cannot lapse into any chance thing you may name other than itself, but only into a more or less complete 'want of culture.'

And it is the same with everything else; for structural combinations no less than isolated qualities obey the same law; only that we fail to note it because we have not specific names for the several 'absences of structure' corresponding to different structural forms. But, all the same, anything that is articulated must rise out of something from which that particular articulation is absent; and if, in its turn, it falls out of articulation it must go back again to the absence of the specific articulation it had. It makes no difference whether we speak of 'harmony' or of arrangement or of combination in this connection; for the principle is clearly the same. And so too with a house or a statue or any such product. For what the house replaces by being made is the unordered relation of the materials to each other; and what passes away in the making of the statue, or any other shapely work, is the unshapeliness of the material; for all such things are constituted either by the formative disposition or the combining of the material or materials.

If then all this is so, it would seem that whenever anything comes into existence or passes out of it, the movement is along the determined line between the terms of some contrast; of (if we start from some intermediate state) the movement is towards one of the extremes. And since the intermediates are compounded in various degrees out of the opposite terms of the contrasted couple (colours, for instance, out of black and white), it follows that all things that come into existence in the course of nature are either opposites themselves or are compounded of opposites.

3. Up to this point, as already indicated, we may claim the pretty general consent of all the systems; for all thinkers posit their elements or 'principles,' as they call them; and, though they give no reasoned account of these 'principles,' nevertheless we find—as though truth itself drove them to it in spite of themselves—that they are talking about contrasted couples.

4. But they differ in the order they follow, some starting from what is most luminous to the intelligence, others from what is more directly cognizable by the senses. Thus some posit 'hot and cold,' or 'wet and dry,' as causes of becoming,

and others 'odd and even,' or 'amity and conflict,' which illustrates what I am saying.

5. Thus they agree in one respect, while differing in others. Their differences are obvious and are universally recognized; but what is not seen so generally is that they are all analogous in so far as they all rest upon the same fundamental conception of antithesis, though some express it in a wider and some in a narrower formula.

To this extent, then, they agree and differ, and do worse or better one than the other; some, as already said, beginning with what is more accessible to intelligence and others with what is more accessible to sense; for the general is approached by the intelligence and the particular by the senses, since the mind grasps the universal principle and the senses the partial application. Thus, 'great and small' are mental conceptions, while 'thick and thin' are sense impressions.

But in any case it is clear that the 'principles' must form a contrasted couple.

The contrariety of first principles according to the ancients.

75(1). After stating the opinions of the ancient philosophers regarding the principles of nature, *he begins to inquire into the truth of the problem*. And first, he investigates it through the method of disputation by proceeding from probables;[a,109] secondly, he determines the truth through the method of demonstration[b,110] (less. 12, n. 1). With regard to the first what he does is two-fold. First, his investigation is concerned with the contrariety of principles, and secondly, with their number (less. 11, n. 1). With regard to the first he makes three points. First, he states the opinion of the ancient philosophers regarding the contrariety of principles; secondly, he gives the reason for this (2); in the third place, he shows how the philosophers were related in regard to the positing of contrary principles (3).[c]

---

[a]Cf. St. Thomas, *In IV Metaph.*, less. 4, nn. 572–575; *In I Post. Anal.*, less. 20, n. 5; *In III Physic.*, less. 8, nn. 1–4; *In I De Caelo*, less. 15, nn. 1–3; *De Trin.*, qu. 4, art. 2; *ibid.*, qu. 6, art. 1; John of St. Thomas, *Curs. Philo.*, Vol. II, Log. 2, qu. 1, art. 5, pp. 277–284; Aristotle, *Topica.*, Bk. I, chap. 1–11.

[b]Cf. Aristotle, *Prior. Anal.*, chap. 1; St. Thomas, *In VI Ethic.*, less. 3, n. 1143; *In I Post. Anal.*, less. 1, nn. 5, 6; F. Edmund Dolan, F.S.C., *op. cit.*, pp. 31–48.

[c]Cf. John of St. Thomas, *Curs. Phil.*, Vol. II, pp. 34–48. This is a good summary of the first ten lessons of *In I Physic.*

76(2). First he says, therefore, that all the ancient philosophers stated contrariety in principles.[111] And this is clarified by three opinions of the philosophers. For some said that the whole universe is one immobile being. Of these, Parmenides said that all things are rationally[a] one but sensibly many[b] and to the extent that they are many, he stated contrary principles in them, viz., hot and cold, and he attributed hot to fire, but cold to earth.[c] But the second opinion was

---

[a]"secundum rationem," i.e., in meaning or definition or as known by the intellect.

[b]Cf. Parmenides, part of frag. 7–8: "...At this point I cease my reliable theory (Logos) and thought, concerning Truth; from here onwards you must learn the opinions of mortals, listening to the deceptive order of my words.

They have established (the custom of) naming two forms, one of which ought not to be (mentioned): that is where they have gone astray. They have distinguished them as opposite in form, and have marked them off from another by giving them different signs; on one side the flaming fire in the heavens, mild, very light (in weight), the same as itself in every direction, and not the same as the other. This (other) also is by itself and opposite; dark Night, a dense and heavy body. This world-order I describe to you throughout as it appears with all its phenomena, in order that no intellect of mortal men may outstrip you." Cf. also frag. 9: "But since all things are named Light and Night, and names have been given to each class of things according to the power of one or of the other (Light or Night), everything is full equally of Light and invisible Night, as both are equal, because to neither of them belongs any share (of the other)." Freeman, *op. cit.*, pp. 44–45. This last parenthesis is from the original text. Cf. also Jean Beaufret, *Le Poème de Parménide* (Paris: Presses Universitaires de Paris, 1955), pp. 17–73; A. E. Taylor, *The Parmenides of Plato*, pp. 14–28; Kathleen Freeman, *Pre-Socratic Philosophers* (Oxford: Basil Blackwell, 1949), pp. 140–152; Cornford, *Plato and Parmenides*, pp. 28–52.

[c]Anaxagoras also attributed cold to earth. Frag. 15: "The dense and moist and cold and dark (Elements) collected here, where now is Earth, and the rare and hot and dry went outwards to the furthest part of the Aether." Frag. 16: "From these, while they are separating off, Earth solidifies; for from the clouds, water is separated off, and from the water, earth, and from the earth, stones are solidified

that of the natural philosophers who stated one material mobile principle,[112] and they said that other things were made from this on the basis of rarity and density, and thus they said that the rare and the dense were principles. The third opinion is that of those who stated many principles. Among them, Democritus held that all things came to be from indivisible bodies, which, while joined to one another, in the very contact left a certain void, and spaces of this kind he called *pores*, as is seen in *On Generation* I.[a,113] In this way, therefore, he said that all bodies were composites of the *firm* and the *empty*, i.e., of plenum and void. And so, he said that the plenum and the void are principles of nature. But he linked the plenum to being and the void to non-being. Likewise, although all indivisible bodies were of one nature, yet he said different things were constituted from these in virtue of the difference of shape, position and order.[b] Thus he stated principles to be contraries in the genus of position, viz., above and below, before and behind, and contraries in the genus of shape, viz., straight, angular or curved, and, likewise, contraries in the genus of order, as prior and posterior, but no mention of these is made in the test because they are obvious. And so, he concludes as if by induction that all philosophers stated principles to be contraries in some way. But he did not mention the opinion of Anaxagoras and Empedocles, because he had explained these in more

---

by the cold; and these rush outward rather than the water." Freeman, *Ancilla to the Pre-Socratic Philosophers*, p. 85. Cf. St. Thomas, *In I Metaph.*, less. 9, nn. 138–139.

[a]St. Thomas, Bk. I, less. 22. We are told in the Leonine edition that this less. belongs to that part of Bk. I which is by an unknown author and not by St. Thomas; cf. *ibid.*, less. 3.

[b]Cf. St. Thomas, *In I Metaph.*, less. 7, esp. n. 116.

detail above.[a] And yet, even these latter stated that there was contrariety in principles in some way, since they said that all things come to be by congregation and segregation,[b] which belong to the genus of rare and dense.

77(3). Then, when he says: (2), he states a probable argument to show that first principles are contraries, and it is as follows: There seem to be three factors that concern the nature of principles. First, they are not from others; secondly, they are not from each other; thirdly, all other things are from them. But these three points belong to the first contraries. The first contraries are, therefore, principles. Moreover, to understand why he calls them *first contraries*, we ought to take into consideration that there are some contraries which find their cause in other contraries,[114] as sweet and bitter are caused from humid and dry, and hot and

---

[a]Less. 8, nn. 4–5.

[b]Cf. Empedocles, frag. 35: "But I will go back to the path of song which I formerly laid down, drawing one argument from another: that (path which shows how) when Hate has reached the bottommost abyss of the eddy, and when Love reaches the middle of the whirl, then in it (the whirl) all these things come together so as to be One—not all at once, but voluntarily uniting, some from one quarter, others from another. And as they mixed, there poured forth countless races of mortals. But many things stand unmixed side by side with the things mixing—all those which Hate (still) aloft checked, since it had not yet faultlessly withdrawn from the Whole to the outermost limits of the circle, but was remaining in some places, and in other places departing from the limbs (of the Sphere). But in so far as it went on quietly streaming out, to the same extent there was entering a benevolent immortal in rush of faultless Love. And swiftly those things became mortal which previously had experienced immortality, and things formerly unmixed became mixed, changing their paths. And as they mixed, there poured forth countless races of mortals, equipped with forms of every sort, a marvel to behold." Freeman, *op. cit.*, pp. 56–57. Cf. also St. Thomas, *In I Metaph.*, less. 6, esp. n. 111.

cold. Now, we cannot proceed in this way to infinity,[115] but we must arrive at some contraries, which do not find their cause in other contraries, and these he calls *first contraries*.[a] The three foregoing conditions of principles belong, therefore, to these first contraries. From the fact that they are *first*, it is clear that they are not from others. But from the fact that they are *contraries*, it is clear that they are not from each other, for, even though the cold comes to be from the hot to the extent that what is first hot afterwards becomes cold, yet coldness itself never comes to be from heat, as will be said later.[b] But we must further investigate the third point, viz., *how all things come to be from contraries*.

78(4). In order to elucidate this, he, therefore, first establishes as a premise that neither action nor passion can occur among *contingent beings*,[116] i.e., among those that merely happen to be together. Nor can they occur among *contingent beings* in the sense of among any indeterminate beings whatsoever.[c,117] Nor does anything whatsoever come to be from anything whatsoever, as Anaxagoras said, unless, perchance, incidentally.

And this is made clear first in simple things, for white does not come to be from musical unless, perchance, incidentally, to the extent that white or black is an accident to someone musical. But white comes to be in the full sense from non-white, and not from any non-white whatsoever,[d,118] but from non-white that is

---

[a]Cf. St. Thomas, *In II De Anima*, less. 21, nn. 514–524.

[b]Less. 11, nn. 8–9.

[c]Cf. St. Thomas, *In VI Metaph.*, less. 2, nn. 1184–1185.

[d]Cf. *ibid.*, Bk. X, less. 6, n. 2038.

black or an intermediate color. And in the same way musical comes to be in the full sense from non-musical, and not from any non-musical whatsoever, but from the opposite which is called unmusical, i.e., what by its nature[a] is meant to have music but does not have it, or from whatever is intermediate between them.[119] And by the same reasoning, something is not primarily and strictly corrupted into any contingent being whatsoever, e.g., white is not corrupted into musical unless by accident, but it is corrupted in the full sense into non-white, and not into any non-white whatsoever, but into black or into an intermediate color.[120] And he says the same about the corruption of musical and of other like things. And the reason for this is that whatever comes to be and is corrupted is not before it comes to be and does not exist after it is corrupted. And so, that which something comes to be in the full sense, and into which something is substantially corrupted, must be such that in its meaning it includes the non-being of what comes to be or is corrupted.[121]

And in a similar way he makes this clear in the case of composites. And he says that the case of composites is similar to that of simple things, but it is less evident in composites, because the opposite of composites are not named, as are the opposites of simple things, for the opposite of house has no name, but the opposite of white does have a name. Thus, if they were reduced to some named thing, it would be obvious. For every composite consists of some harmony; but *harmony* comes to be from discord, and discord from harmony; and likewise, harmony is corrupted into discord, but not into any discord whatsoever, but into its opposite.[b] Moreover, discord can be spoken of either on the basis of order only

---

[a]"natum."

[b]Cf. St. Thomas, *Cont. Gent.*, Bk. II, chap. 64; *In I Ethic.*, less. 6, n. 10; *ibid.*, less. 1, n. 5.

or on the basis of composition. For there is one type of whole that consists in the harmony of order, e.g., an army, but another that consists in the harmony of composition,[122] e.g., a house. And this same reasoning holds for both. And it is evident that all composites come to be similar from what is not composite, as a house comes to be from what is not composite, and shape from the unshaped; and in all these things attention is given only to order and composition.

So, therefore, as if by induction, it is evident that all things which come to be or are corrupted, come to be from contraries or from their intermediaries or are corrupted into them. But intermediaries of contraries come to be from contraries as intermediate colors come to be from white and black. And so, he concludes that all beings that come to be in nature, either themselves are contraries, as white and black, or come to be from contraries, as what are intermediaries of contraries. And this is the principal point that he intends to conclude, viz., that *all things come to be from contraries, which was the third condition prior to principles.*

79(5). Then, when he says: (3), the Philosopher shows here what the positions of the philosophers were with regard to stating that principles are contraries.[a] First, he shows what their positions were with regard to the motive of their statements, and secondly, he shows what their positions were with regard to the statements themselves (4). First, he says, therefore, that many of the philosophers, as was said above,[b] pursued the truth up to this point, viz., that they stated principles to be contraries. But, although some spoke truly, yet, they did not

---

[a]St. Thomas, *In I De Caelo*, less. 6, n. 10.

[b]N. 2.

speak as if moved by some argument,[a] but as if compelled by truth itself.[123] For *truth* is the good of the intellect to which it is naturally ordered. And so, just as beings that lack knowledge are moved to their ends without reason, so, sometimes man's intellect tends to truth by a certain natural inclination, although it does not perceive the reason for the truth.[b,124]

80(6). Then, when he says: (4), he shows the position of the foregoing philosophers in the statements that they made. Concerning this he makes two points. First, he shows how they differ in stating that principles are contraries, and secondly, he shows how they differ and agree at the same time (5).

He says, therefore, that the philosophers who said that principles are contraries, differed in two ways. First, they differed because some of them who spoke on the basis of reason took *prior contraries* as principles, but others who manifested less foresight in their considerations, took *posterior contraries* as principles. And of those who took *prior contraries*, some directed their attention to those that are more known on the basis of reason,[c,125] but certain others to those that are more known sensibly. It can even be said that the second difference is the reason for the first, for those contraries that are more known on the basis of reason are *prior absolutely*, but those that are more known sensibly are *posterior absolutely*

---

[a]"aliqua ratione."

[b]On certitude cf.: St. Thomas, *Summa Theol.*, IIa IIae, qu. 4, art. 8; *In VI Ethic.*, less. 3, n. 1143; *In I Post. Anal.*, less. 25, 44; *In II Post. Anal.*, less. 1, n. 6; *ibid.*, less. 20; *In III De Anima*, less. 4, nn. 630–635, esp. 632. One might also consult: *ibid.*, less. 5, n. 649; Aristotle, *Topica*, Bk. I, chap. 10, 11.

[c]"secundum rationem," i.e., after they reasoned them out.

and prior in relation to us. But it is clear that principles must be first. And so, those who judged that that is prior which is more known on the basis of reason,[a] stated contrary principles that are prior absolutely, but those who judged that that is prior which is more known sensibly stated principles that are posterior absolutely. And so, some stated as first principles the hot and the cold, but others stated humid and dry, and both of these sets of contraries are more known sensibly. Yet, hot and cold, which are active qualities, are prior to humid and dry, which are passive qualities, because the active is naturally prior to the passive. But others stated principles that are more known on the basis of reason. Some of these said that *even* and *odd*[b] were principles. These were the Pythagoreans,[c] who thought that

---

[a]"rationi."

[b]Cf. St. Thomas, *In I Metaph.*, Bk. I, less. 8, esp. n. 125; cf. also Philolaus of Tarentum, frag. 4: "Actually, everything that can be known has a Number; for it is impossible to grasp anything with the mind or to recognise it without this (Number)." Cf. also frag. 5: "Actually, Number has two distinct forms, odd and even, and a third compounded of both, the even-odd; each of these two forms has many aspects, which each separate object demonstrates in itself." Freeman, *op. cit.*, p. 74. "Philolaus of Tarentum was active in the latter half of the fifth century B.C.

He was said to have written one book, which was the first published account of Pythagoreanism. The fragments attributed to him are in Doric dialect. Their genuineness has been disputed by modern scholars, probably without justification. This work was usually given the title *On the Universe*. Another work entitled *Bacchae* was sometimes attributed to him." *Ibid.*, p. 73.

[c]"Pythagorean School. Accounts of Pythagoreanism derive mostly from Aristotle and the Peripatetic School (Aristoxenus, Theophrastus, Eudemus), and from the Neo-Platonists (Porphyry, Iamblichus, Proclus, Simplicius); there are also extracts and references in the compilers (Diogenes Laertius, Stobaeus) and the lexicographers. But these accounts are usually referred to the Pythagorean School in general, not to any particular member. It was a rule that discoveries were referred to Pythagoras himself; and Pythagoras left no writings." *Ibid.*, p. 82.

the substance of everything is numbers,[a,126] and that all things are composed of *even* and *odd*, as of form and matter, for they attributed infinity and otherness to *even* because of its divisibility, but they attributed finiteness and identity to *odd* because of its indivision. But others, viz., the followers of Empedocles, said that the causes of generation and corruption were *discord* and *harmony*,[b,127] which are

---

[a]Cf. Philolaus of Tarentum, frag. 11: "One must study the activities and the essence of Number in accordance with the power existing in the Decad (Ten-ness); for it (the Decad) is great, complete, all-achieving, and the origin of divine and human life and its Leader; it shares...The power also of the Decad. Without this, all things are unlimited, obscure and indiscernible.

For the nature of Number is the cause of recognition, able to give guidance and teaching to every man in what is puzzling and unknown. For none of existing things would be clear to anyone, either in themselves or in their relationship to one another, unless there existed Number and its essence. But in fact Number, fitting all things into the soul through sense-perception, makes them recognisable and comparable with one another as is provided by the nature of the Gnômôn, in that Number gives them body and divides the different relationships of things, whether they be Non-Limited or Limiting, into their separate groups.

And you may see the nature of Number and its power at work not only in supernatural and divine existences but also in all human activities and words everywhere, both throughout all technical production and also in music.

The nature of Number and Harmony admits of no Falsehood; for this is unrelated to them. Falsehood and Envy belong to the nature of the Non-Limited and the Unintelligent and the Irrational.

Falsehood can in no way breathe on Number; for Falsehood is inimical and hostile to its nature, whereas Truth is related to and in close natural union with the race of Number." *Ibid.*, p. 75. Cf. also J. Burnet, *Greek Philosophy from Thales to Plato* (London: Macmillan, 1914), pp. 51–54, 83, 85, 89; J. E. Raven, *Pythagoreans and Eleatics* (Cambridge: The University Press, 1948), entire book but esp. pp. 112–187.

[b]Cf. Empedocles, frag. 20: "This process is clearly to be seen throughout the mass of mortal limbs; sometimes through Love all the limbs which the body has as its lot come together into One, in the prime of flourishing life; at another time again, sundered by evil feuds, they wander severally by the breakers of the shore

also more known on the basis of reason.[128] And so, it is clear that in these statements the difference mentioned above is apparent.

81(7). Then, when he says: (5), he shows how there is also a certain agreement in the differences in the foregoing opinions. The fact of this agreement proves that in one way the ancient philosophers stated the same principles, and in another way, the stated different ones. They were *different*, indeed, to the extent

---

of life. Likewise too with shrub-plans and fish in their watery dwelling, and beasts with mountain lairs and diver-birds that travel on wings." Freeman, *op. cit.*, p. 54. Cf. also frag. 21: "But come, observe the following witness to my previous discourse, lest in my former statements there was any substance of which the form was missing. Observe the sun, bright to see and hot everywhere, and all the immortal things (heavenly bodies) drenched with its heat and brilliant light; and (observe) the rain, dark and chill over everything; and from the Earth issue forth things based on the soil and solid. But in (the reign of) Wrath they are all different in form and separate, while in (the reign of) Love they come together and long for one another. For from these (Elements) come all things that were and are and will be; and trees spring up, and men and women, and beasts and birds and water-nurtured fish, and even the long-lived gods who are highest in honour. For these (Elements) alone exist, but by running through one another they become different; to such a degree does mixing change them." *Ibid.*, pp. 54–55; and cf. also frag. 22: "For all these things—beaming Sun and Earth and Heaven and Sea—are connected in harmony with their own parts: all those (parts) which have been sundered from them and exist in mortal limbs. Similarly all those things which are more suitable for mixture are made like one another and united in affection by Aphrodite. But those things which differ most from one another in origin and mixture and the forms in which they are moulded are completely unaccustomed to combine, and are very baneful because of the commands of Hate, in that Hate has wrought their origin." *Ibid.*, p. 55. Cf. also St. Thomas, *In I Metaph.*, less. 6; *In I De Gen. et Corr.*, less. 2; *In II De Gen. et Corr.*, less. 7; Jean Zafiropulo, *Empédocle d'Agrigente* (Paris: Société d'édition "Les Belles Lettres," 1953), pp. 95–148; William Ellery Leonard, *The Fragments of Empedocles* (Chicago: The Open Court Publishing Company, 1908), pp. 4–12.

that different philosophers assumed different contraries, as was said,[a] but the same *analogously*, i.e., according to a proportion, because the principles taken by all have the same proportion.

And he says this is true for three reasons. The first reason is that whatever principles they take are related to one another as contraries. And he says that all understand principles *from the same reciprocal order*,[b] viz., from the order of contraries, for all accept contraries as principles, yet they take different contraries. It is not strange that different principles are derived from one reciprocal order of contraries, because among contraries some contain, as the prior and the more common ones, and others are contained, as the posterior and less common ones. This is, therefore, the one way in which they speak similarly, viz., in that they all understand principles as based on the order of contraries.

Another way in which they agree analogously is that no matter what principles they accepted, one of these principles was considered[c] as better and the other as worse.[129] For example, harmony or the full or the hot were considered better as principles, but discord or the void or the cold were considered worse as principles. And the same is true with regard to the consideration of other contraries. And this is so, because the other of the contraries always has privation connected with it, for the principle of contrariety is the opposition of privation and possession,[d] as is said in *Metaphysics* X.[e,130]

---

[a]N. 6, par. 2.

[b]"Coordinatione," i.e., the relationship that there is in order.

[c]"se habet."

[d]"habitus."

[e]St. Thomas, *In X Metaph.*, less. 6, nn. 2036–2037.

The third way in which they agree analogously is this, viz., that all assume principles that are better known.[a,131] Some, however, take those that are better known on the basis of reason, and others take those better known sensibly, for, since reason is universal, and sense particular, universals, as large and small, are more known on the basis of reason, but singulars, e.g., rare and dense, which are less common, are more known sensibly. And thus as an epilogue, he finally comes to the conclusion which he principally intended, viz., that of showing that *principles are contraries.*

---

[a]"notiora."

LESSON ELEVEN[a]

1. The ground might now seem to be clear for discussing whether the ultimate principles of Nature are two, or three, or some greater but limited number.

2. A single principle will not do, for we have established antithesis as an ultimate constituent in Nature, and antithesis involves duality.

3a. Nor can the factors of Nature be unlimited, or Nature could not be made the object of knowledge.

3b. Now, as far as antithesis goes, two principles would be enough, for every defined class comes under one general antithesis, and the whole sum of 'things that exist in Nature,' as such, forms a defined class. Since, then, one antithesis will suffice, it is better not to go beyond it,

4. for the more limited, if adequate, is always preferable, as we saw in the case of Empedocles, who claims to get everything out of his four substances that Anaxagoras claims to get out of his unlimited number.

5. Or we may put it that some antitheses are more general than others, and some are derived from others. Such subordinate antitheses are exemplified by 'sweet and bitter' or 'black and white.' But the fundamental principles must be in evidence in *every* case. Obviously, then, since antithesis is fundamental and implies duality, the ultimate principles, though limited in number, must be more than one.

6. But, granting them to be limited, there are reasons for supposing them to be more than two. For (1) a man might well be at a loss to conceive how 'density' itself, for instance, could possibly make 'rarity's' self into anything, or 'rarity' density's. And so too with any other antithesis; for, I suppose, we are not asked to think of 'amity' drawing 'hostility' closer together and thereby constructing

---

[a]Includes 11. 189 a 11—189 b 29, Wicksteed and Cornford, *op. cit.*, pp. 59–67.

something out of it, nor *vice versa,* rather than both of them acting upon a third something. (2) And indeed some thinkers have assumed more than one such principle out of which to construct the nature of things.

7. (3) Besides, one might encounter another difficulty in supposing there to be nothing in Nature underlying the antithetical couple; for we never see the antithetical principles, by themselves, constituting the substantive existence of anything we know in Nature, and a true principle cannot be the mere attribute of something else, for the subject to which it was attributed would be a principle anterior to it, inasmuch as it would be presupposed in the very fact of having this attribute predicted of it.

8. (4) Nor can we make 'natural existence' the term of an antithesis, for things move from one term of an antithesis to the other, and since nothing 'exists in Nature' that is contrasted with 'natural existence,' how could any such existent move out of what is not there for it to move out of? Or how could such a non-existent be a presupposition of the existent?

9. Thus, if our former insistence on the two terms of some antithesis being principles is sound, and if we are now convinced that these antithetical principles need something to work on, and if we are to preserve both these conclusions, must we not necessarily posit a third principle as the subject on which the antithetical principles act? And it is this third principle that those physicists are really thinking about who say that the universe has but one single constituent, water or fire or some intermediate substance. Of these the hypothesis of an intermediate substance seems best, for fire, earth, air, and water, are themselves implicated with contrasted properties; so that there is reason in distinguishing the universal material element from all of them; though some have identified it with air, inasmuch as the special characteristics of air are less obtrusive on the senses than those of the others, water coming next in this respect.

10. But the Physicists, each and all, mould this universal subject matter by such antithetical principles as density and rarity, or 'more or less,' all which couples may be reduced to 'excess and defect' as already pointed out. In reality, this doctrine, that the 'one' and 'excess and defect' are the principles of all that is, turns out to have been in possession of men's minds from of old; but not always in the same way, for the earlier thinkers made the 'one' the receptive subject and the contrasted 'two' the active agents, whereas more recently the 'two' have

sometimes been regarded as the subject passively acted upon and the 'one' as the agent.

11. So from these and other considerations it appears that there is much to be said for assuming three principles, but not for going beyond three. For one passive principle is enough. And if there were supposed to be four altogether, and there were to constitute two couples, each couple would demand a subject of its own to act upon, apart from the other couple; or, if one were derivative from the other, it would not count among the 'principles,' for it would not be primary.

12. Nor could there be two primary antitheses; for 'natural existence' groups 'all that is in Nature' under a single aspect, so that the supposed pairs of principles could only differ from each other in priority, and could not belong to different series; for there is never more than one general antithesis common to one group, and it seems that all antitheses can be brought under one general antithesis.

It is clear, then, that there must be more than one element of principle, and that there cannot be more than two or three. But, within these limits, the decision as between two and three presents great difficulties.

There are three principles of natural beings, no more and no less.

82(1). Now that the Philosopher has investigated the contrariety of principles, *here he begins to investigate their number*. With regard to this what he does is threefold. First, he proposes the question; secondly, he excludes what does not enter into the inquiry (2), and thirdly he pursues the inquiry (6). He says first, therefore, that after the inquiry about the contrariety of principles, the investigation of their number, viz., whether they are two or three or more, follows as a natural consequence.

83(2). Then, when he says: (2), he excludes what does not enter into the inquiry. And he says first that there is not one principle only; secondly, he says that they are not infinite (3). He says in the first place, therefore, that it is impossible that there be but one principle, for it has been shown[a] that principles are contraries, and contraries are not one only, for nothing is contrary to itself; therefore, principles are not one only.

84(3). Then, when he says: (3), he shows by four arguments that principles are not infinite. The first of these is as follows: The infinite, as such, is unknown. If, therefore, principles are infinite, they must be unknown. But when the

---

[a]Less. 10.

principles are unknown, what is from them is unknown. It follows, therefore, that nothing in the world can be known.

85(4). He states the second argument (3a). It is as follows: Principles must be primary contraries, as was shown above.[a] Primary contraries are, however, of the first genus. This is substance. And substance, since it is one genus, has one primary contrariety, for the primary contrariety of any genus whatsoever is that of the first differences that divide the genus.[b,132] Principles, therefore, are not infinite.

86(5). He states the third argument (4), and it is as follows: What can come into being by finite principles should be said to come into being by finite principles rather than by infinite ones. But the reason[c] for all things that come to be naturally[d] is assigned, according to Empedocles, to finite principles, and by Anaxagoras, to infinite principles. We must not, therefore, say that principles are infinite.

87(6). He states the fourth argument (5), and it is as follows: Principles are contraries. If, therefore, principles are infinite, all contraries must be principles. But not all contraries are principles. This is obvious for two reasons, first, indeed,

---

[a]Less. 10, nn. 3–4.

[b]Cf. Aristotle, *Categoriae*, chap. 5, 6.

[c]"ratio."

[d]"secundum naturam."

because principles must be primary contraries and not all contraries are primary, since some are prior to others, and secondly, because principles ought not be free from each other, as was said above.[a] Some contraries, however, come into being from each other, as do sweet and bitter, and white and black. Principles are, therefore, not infinite. And so, he finally concludes that principles are neither one only nor infinite.

88(7). Moreover, we should consider that here the Philosopher proceeds disputatively from probable premises. This is why he assumes what is seen by many, and what cannot be wholly false but is partially true. It is true in some way, therefore, that contraries come into being from one another, as was said above,[b] viz., if the subject is taken with the contraries, because what is white afterwards becomes black. Yet, whiteness itself is not converted into blackness. But some ancient philosophers said that first contraries do not come into being from one another, even if we take the subject along with these. And so, Empedocles denied that elements came into being from one another.[c] And, therefore, Aristotle

---

[a]Less. 10, first part of n. 3.

[b]*Ibid.*, n. 4.

[c]Cf. *infra*, Bk. II, nn. 169, 202; cf. also Empedocles, frag. 37: "(Fire increases fire), Earth increases its own substance, Aether (increases) Aether." Freeman, *op. cit.*, p. 57. For Empedocles, the elements did not come from nothing, nor are they corrupted into nothing. Rather, beings were made from an existing something according to a determined mixture. This is clear in the ff. fragments: frag. 9: "But men, when these (the Elements) have been mixed in the form of a man and come into the light, or in the form of a species of wild animals, or plants, or birds, they say that this has 'come into being'; and when they separate, this men call sad fate (death). The terms that Right demands they do not use; but through custom I myself also apply these names." Frag. 11: "Fools!—for

deliberately does not say here that the hot comes to be from the cold, but the sweet comes to be from the bitter and the white from the black.

89(8). Then, when he says: (6), he continues with what was in question, viz., the number of principles. And with regard to this, what he does is threefold. First, he shows that there are not two principles only, but three, and secondly, he shows that there are no more than three (11).

With regard to the first point, what he does is twofold. First, he shows by arguments that there are not two principles only, but that we must add a third. And secondly, he shows that the ancient philosophers also agreed in this (9).

90(9). With regard to the first point, he states three arguments. Thus, he says first that it has been shown[a] that principles are contraries, and that for this reason there cannot be one principle only but there are at least two; and it has also been shown that there are not infinite principles. Since all this has been shown, it now remains for us to consider whether there are two only or more than two. For,

---

they have no long sighted thoughts, since they imagine that what previously did not exist comes into being, or that a thing dies and is utterly destroyed." Frag. 12: "From what in no wise exists, it is impossible for anything to come into being; and for Being to perish completely is incapable of fulfillment and unthinkable; for it will always be there, wherever anyone may place it on any occasion." Frag. 13: "Nor is there any part of the Whole that is empty or overfull." Frag. 14: "No part of the Whole is empty; so whence could anything additional come?" Cf. also frag. 15: "A wise man would not conjecture such things in his heart, namely, that so long as they are alive (which they call Life), they exist, and experience bad and good fortune; but that before mortals were combined (out of the Elements) and after they were dissolved, they are nothing at all." *Ibid.*, pp. 52–53.

[a]Less. 10, nn. 2, 5.

with regard to what was shown above,[a] viz., that principles are contraries, it seems that there are only two principles, because contrariety exists between two extremes. But in this, *someone will falter*,[b] i.e., he will have a doubt, for other things come to be from principles, as was said above.[c] If, however, there are only two contrary principles, it is not clear how all things can come into being from these two, for we cannot say that one of them makes something from the one that is left, for density is not naturally fit to convert rarity into something nor rarity density. And the same is true of any other contrariety, for harmony does not cause discord and make something from it, nor conversely. But both contraries change a third something, which is the subject of both, for the hot does not make coldness itself to be hot, but makes the subject of coldness to be hot, and not conversely. It seems, therefore, that we must state some third thing, which is the subject of the contraries. Whether that subject be one or many does not make any difference as far as the present consideration is concerned, for some have posited many material principles and from these they constructed the nature of beings, for they said that matter alone was the nature of all things, as will be said below in the second book.[d]

91(10). He states the second argument (7), and he says that unless something else is supposed besides the contraries which are stated to be principles, greater doubt than the preceding will arise. A first principle, indeed, cannot be some

---

[a]Less. 10.

[b]"deficiet."

[c]Less. 10, n. 3.

[d]Less. 2, nn. 1–2.

accident said of a subject, for, since the subject is the principle of an accident which is predicated of it, and since it is naturally prior to it, it should follow, if the first principle were an accident predicated of a subject, that it would be the principle of a principle and that it would be something prior to what is first. But if we state contraries alone to be principles, a principle would have to be some accident said of a subject, for there is nothing whose substance is a contrary to another,[a,133] but contrariety exists only among accidents. It remains, therefore, that contraries alone cannot be principles. We should consider, however, that in this argument he uses *predicate* as *accident*, because a predicate designates the form of the subject, but the ancient philosophers believed that all forms were accidents. Here, however, he is proceeding disputatively from probable propositions which were popular among the ancients.

92(11). He states the third argument (8), and it is as follows: Everything that is not a principle must be from principles. If, therefore, contraries alone are principles, it follows, since substance is not the contrary of substance, that substance is from non-substances. And so, what is not substance exists before substance, because what is from other things comes after those things. But this is impossible, because the first genus of being is substance which is being of itself.[b] Thus, it cannot be that contraries alone are principles, but we must posit some other third thing.

---

[a]Cf. Aristotle, *Categoriae*, chap. 5.

[b]"per se."

93(12). Then, when he says: (9), he shows how the statement of the philosophers also agreed with this. And with regard to this what he does is twofold. First, he shows how they stated one material principle, and secondly, how in addition to this they stated two contrary principles (10).

But with regard to the first point, we should consider that in the preceding arguments the Philosopher seemed to be opposed to each opposite viewpoint, as is the custom with disputers. For first,[a] he proved that principles are contraries, and now[b] he has brought forth arguments to prove that contraries are not sufficient for this, viz., that things be generated from them. And because the disputative arguments conclude to truth in some way, although not entirely, he concludes to one truth from both arguments.[134]

And he says that if anyone should consider as true both the prior argument, that states that principles are contraries, and this argument just stated, which proves that contrary principles cannot suffice, he must say, in order to save both arguments, that a third something underlies the contraries, as did those who stated the whole universe to be some one nature. By the term nature they meant matter, as water or fire or air or what is intermediate between them, as vapor or something of that kind.

What is intermediate between them seems preferable. This third thing, indeed, is taken as a subject of the contraries; and in some way as distinct from them.[c,135] And so, what has less contrariety is more fittingly stated to be the third

---

[a]Less. 10, n. 3—end of less.

[b]This less., nn 9–11.

[c]Cf. St. Thomas, *In II De Anima*, less. 22, n. 524; *In V Metaph.*, less. 12, n. 926.

principle in addition to the contraries, for fire, earth, air and water have contrariety joined to[a] them, viz., that of hot and cold, moist and dry.[b] And so, they do not unreasonably make the subject to be something distinct from them, and something in which there is less of the excess of contraries. Next, those who stated air[c] to be

---

[a]"annexam."

[b]Cf. Heraclitus, frag. 126: "Cold things grow hot, hot things grow cold, the wet dries, the parched is moistened." Freeman, *op. cit.*, p. 33; frag. 88: "And what is in us is the same thing: living and dead, awake and sleeping, as well as young and old; for the latter (of each pair of opposites) having changed becomes the former and this again having changed becomes the latter." *Ibid.*, p. 30.

[c]As Diogenes did. "Diogenes of Apollônia (probably on the Black Sea), lived in the latter half of the fifth century B.C.

His longest surviving work was that *On Natural Science*; he also wrote, and mentioned in his main work, separate treatises on Meteorology, and On the Nature of Man; and an attack on the Natural Scientists, whom he called Sophists. Frag. 2: "It seems to me, to sum up the whole matter, that all existing things are created by the alteration of the same thing, and are the same thing. This is very obvious. For if the things now existing in this universe—earth and water and air and fire and all other things which are seen to exist in this world: if any one of these were different in its own (essential) nature, and were not the same thing which was transformed in many ways and changed, in no way could things mix with one another, nor could there to any profit or damage which accrued from one thing to another, nor could any plant grow out of the earth, nor any animal or any other thing come into being, unless it were so compounded as to be the same. But all these things come into being in different forms at different times by changes of the same (substance), and they return to the same." Frag. 3: "Such a distribution would not have been possible without Intelligence, (namely) that all things should have their measure: winter and summer and night and day and rains and winds and periods of fine weather; other things also, if one will study them closely, will be found to have the best possible arrangement." Frag. 4: "Further, in addition to these, there are also the following important indications: men and all other animals live by means of Air, which they breathe in, and this for them is both Soul (Life) and Intelligence, as had been clearly demonstrated in this treatise; and if this is taken from (them), Intelligence also leaves them." Frag. 5: "And it seems to me

principle spoke more correctly, because in air there are fewer sensible contrary qualities. Those who posited water were next. But those who posited fire spoke the least correctly in this regard, because fire has a contrary quality that is most sensible and more active.[a] The reason for this is that in fire there is the excess of warmth. If, however, the elements were compared as to their subtlety, those who stated fire as a principle would seem to have spoken more correctly, as is said

---

that that which has Intelligence is that which is called Air by mankind: and further, that by this all creatures are guided, and that it rules everything; for this in itself seems to me to be God and to reach everywhere and to arrange everything and to be in everything. And there is nothing which has no share of it; but the share of each thing is not the same as that of any other, but on the contrary there are many forms both of the Air itself and of Intelligence; for it is manifold in form: hotter and colder and dryer and wetter and more stationary or having a swifter motion; and there are many other differences inherent in it and infinite (forms) of savour and colour. Also in all animals the Soul is the same thing, (namely) Air, warmer than that outside in which we are, but much colder than that nearer the sun. This degree of warmth is not the same in any of the animals (and indeed, it is not the same among different human beings), but it differs not greatly, but so as to be similar. But in fact, no one thing among things subject to change can possibly be exactly like any other thing, without becoming the same thing. Since therefore change is manifold, animals also are manifold and many, and not like one another either in form or in way of life or in intelligence, because of the large number of (the results of) changes. Nevertheless, all things live, see and hear by the same thing (Air), and all have the rest of Intelligence also from the same." *Ibid.*, pp. 87–88.

[a]For Heraclitus, fire was so active that he gave to it amazing powers. Cf. frag. 64: "The thunder-bolt (i.e., Fire) steers the universe." Frag. 66: "Fire, having come upon them, will judge and seize upon (condemn) all things." *Ibid.*, p. 29; frag. 76: "Fire lives the death of earth, and air lives the death of fire; water lives the death of air, earth that of water." *Ibid.*, p. 30.

elsewhere, because whatever is more subtle seems to be simpler and prior.[a,136] Fo this reason, no one stated the earth as a principle because of its bulkiness.

94(13). Then, when he says: (10), he shows how with one materia principle they stated contrary principles. And he says that all who stated on material principle said that it was shaped or formed out of some sort of[b] contraries as rarity and density, which are reduced to the great and the small, and to exces and defect. And so, what Plato maintained, viz., that the one and the great and th small are the principles of things,[137] was also the opinion of the ancient natura philosophers, but in a different way. For the ancient philosophers who considerec one matter to be varied by different forms, stated two principles on the side o form, which is the active principle, and one on the side of matter, which is th passive principle.[138] The Platonists, however, considering how in one species many individual beings are distinguished on the basis of the division of matter, stated one principle on the side of form, which is the active principle, and two on the side o matter, which is the passive principle. And so, he concludes his main point, viz., that to those considering the preceding and similar arguments, it will seem reasonable that there are three principles of nature. And this he says, pointing out that he had preceded from probable premises.

---

[a]Cf. Aristotle, *De Caelo*, Bk. III, chap. 5, nn. 1 ff.; St. Thomas, *In I Metaph.*, less. 12; *In XI Metaph.*, less. 1, nn. 2171–2172; *In III De Caelo*, less. 9, entire less.

[b]"quibusdam."

95(14). Then, when he says: (11), he shows by two arguments that there are not more than three principles. The first of these arguments follows. It is superfluous that something come to be through many principles which can come to be through fewer principles. But the whole generation of natural things can be accomplished by stating one material and two formal principles, because for something to be acted upon one material principle suffices. But if there were four contrary principles and two primary contrarieties, each contrariety would have to have a different subject, because one subject seems to have primarily one contrariety.[139] And in this way, if, when two contraries and one subject have posited, things can come to be from one another, it seems superfluous to state another contrariety. We must, therefore, state no more than three principles.

96(15). He states the second argument[140] (12). And it is as follows. If there are more than three principles, there must be more primary contrarieties. But this is impossible, because primary contrariety seems to belong in the first genus, which is one, i.e., substance. And so, all contraries which are in the genus of substance do not differ in genus, but are related as prior and posterior.[141] The reason for this is that in one genus there is only one contrariety, viz., the first one, in that all the other contrarieties seem to be reduced to the first one, for there are some first contrary differences that divide the genus. Thus, it seems that there are no more than three principles.

Moreover, we must consider that each argument has been stated as probable, viz., both that there is no contrariety in substances, and that there is one prime contrariety in substances. For, if we take a thing which is a substance, nothing is contrary to it, but if we take the formal differences within the genus of substance there is contrariety in them.

97(16). As an epilogue, he finally concludes that there is neither one principle only nor more than two or three. But which of these is true, viz. whether they are two only or three arouses much doubt, as is clear from what has been said.

1. In advancing now to the formulation of a positive theory, let us begin with the general conception of 'change' (that is to say, of things 'coming into existence' altogether, or 'becoming this or that' in particular which they were not before). For the natural order of exposition, as we have seen, is to start from the general principle and proceed to the special applications.

2. Note, then, that in speaking of one thing becoming another, or one thing coming out of, or in the place of, another, we may use either (1) simple or (2) complex terms. I mean that we can say either (1) that a 'man' becomes cultured, or that the 'uncultured' in him is replaced by culture, or (2) that the 'uncultured man' becomes a 'cultivated man.' In this case (1) the 'man' (who acquires culture), and his state of 'unculture' (which is replaced by culture) and the 'culture' itself (which was not, but has 'come to be') are all what I call 'simple' terms; whereas (2) both the 'uncultivated man,' who became something he was not, and the 'cultivated man' that he became, are what I call 'composite' terms.

3. And note that in some of these cases we can say, not only that a thing 'becomes so-and-so,' but also that it does *'from being* so-and-so'; e.g. a man becomes cultivated from being uncultivated. But we cannot use this expression in all cases; for he does not become cultivated *'from being* a man'; on the contrary, he becomes a cultivated *man*.

4. And of the two simple terms, 'man' and 'uncultivated' (both of which we said 'became' something), one (the 'man') persists when he has become a cultivated man; but the other (the 'uncultivated' or 'non-cultivated' in him) does not persist either in the simple 'cultivated' which we say it has 'become' or in the composite 'cultivated man.'

---

[a]Includes 11. 189 b 30—190 b 16, Wicksteed and Cornford, *op. cit.*, pp. 71–79.

5. Observing these distinctions, we may reach a principle of universal application (if we 'observingly distil it out,' as they say), namely that in all cases of becoming there must always be a subject—the thing which becomes or changes and this subject, though constituting a unit, may be analyzed into two concepts and expressed in two terms with different definitions; for the definition of 'man' is distinct from the definition of 'uncultivated.'

6. And the one persists while the other disappears—the one that persists being the one that is *not* embraced in antithesis; for it is the 'man' that persists, and neither the simple 'cultured' or 'uncultivated' nor the composite 'uncultivated man.'

7. When we speak of something 'becoming *from* or *out of*' whatever it may be (rather than of its 'becoming so-and-so'), we generally mean by the thing *from* or *out of* which the becoming takes place, the non-persistent term of aspect: thus, we speak of becoming cultivated from being uncultivated, not from being a man. Still the expression 'out of' is used sometimes of the factor which persists: we say a statue is made out of bronze, not that the bronze *becomes* a statue (in the sense of ceasing to be bronze). When, however, the thing *from* or *out of* which the becoming occurs is the contrasted, non-persistent term, both expressions are used: we can say of a thing, e.g., 'the uncultivated' (man), either that he 'becomes this' (cultivated) or that he 'becomes this (cultivated) *from being* that (uncultivated).' Hence it is the same with the composite terms: we say that the 'uncultivated man' becomes cultivated, and also that he becomes so '*from being* an uncultivated man.'

8. But there is (in Greek) a further ambiguity; for the same word (*gignesthai*) is employed either of a thing 'coming to be' in the absolute sense of 'coming into existence,' or in the sense of 'coming to be this or that' which it was not before; and it is only of a concrete thing, as such, that we can speak of its 'coming to be' in the full sense of coming into existence. Now in all other cases of change, whether of quantity or quality or relation or time or place, it is obvious that there must be some underlying subject which undergoes the change, since it is only a concrete something that can have that 'substantive existence,' the characteristic of which is that it can itself be predicated of no other subject, but is itself the subject of which all the other categories are predicated; but on further consideration it will be equally obvious that a substance also, or anything, whether natural or artificial, that exists independently, proceeds from something that may be regarded as the subject of that change which results in its coming into being; for

in every case there is something already there, out of which the resultant thing comes; for instance the sperm of a plant or animal.

9. The processes by which things 'come into existence' in this absolute sense may be divided into (1) change of shape, or with the statue made of bronze, or (2) additions, as in things that grow, or (3) subtractions, as when a block of marble is chipped into a Hermes, or (4) combination, as in building a house, or (5) such modifications as affect the properties of the material itself. Clearly, then, all the processes that result in anything 'coming to exist' in this absolute sense start with some subject that is already there to undergo the process.

10. From all this it is clear that anything that 'becomes' is always complex: there is (1) something that begins to exist (the new element of form), and (2) something that 'comes to be this' (comes to have this form); and this second thing may be regarded under two aspects—as the subject which persists, or as the contrasted qualification (which the new form will replace). For instance, in the uncultivated man who becomes cultivated, 'uncultivated' is the contrasted qualification, 'man,' the subject; or, when the statue is made, the contrasted qualification is the unshapeliness, formlessness, want of purposeful arrangement; the subject is the bronze or marble or gold.

[illegible] case that [illegible] something already there, out of which [illegible] [illegible] for instance the [illegible] of a plant or animal.

9. The [illegible] by which [illegible] comes into existence [illegible] this [illegible] [illegible] (1) [illegible] or [illegible] (2) [illegible] things [illegible] (3) [illegible] [illegible] and [illegible] [illegible] a building or house, or [illegible] [illegible] [illegible] [illegible] [illegible] [illegible] [illegible] with some subject that was [illegible] there beforehand.

10. [illegible] is clear from [illegible] of [illegible] or [illegible] [illegible]

LESSON TWELVE

In the becoming of any natural being there are three principles: subject, term and opposite.

98(1). After the Philosopher has proceeded disputatively to investigate the number of principles, *he begins to determine the truth.* And this procedure is divided into two parts. In the first, he determines the truth, and in the second, by reason of the truth determined he excludes the doubts and errors of the ancients (less. 14, n. 1). The first of these is divided into two parts. In the first part, he shows that there are three factors in any natural thing that comes into being; secondly, he shows from this that these three are principles (less. 13, n. 1). With regard to the first, what he does is twofold. First, he tells what his intention is, and secondly, he follows through what he intended (2).

99(2). Because he had said above[a] that it is quite doubtful whether there are only two principles of nature or three, he concludes, therefore, that we must speak about this first by considering generation or coming into being in general, and then by considering all the species of change.[b] For in any change, there is a certain *becoming*, just as in what is altered from white to black, non-white comes into being from white, and black comes into being from non-black. And the same is

---

[a]Less. 11, n. 16.

[b]Progress to concretion.

true in other changes. He gives as the reason for this order the fact that we must speak of general things first, and then afterwards speculate about what is proper to each thing, as was said in the beginning of this book.[a]

100(3). Then, when he says: (2), he deals with the point in question.[b] And he deals with this in two ways. First, he sets forth some premises that are necessary in order to clarify the point in question; secondly, he clarifies it (5). And again with regard to the first, what he does is twofold. First, he prefaces a certain division, and secondly, he shows the differences between the parts of the division (3).

101(4). And so, he says in the first place that, since in any becoming at all we say that *one thing* comes into being from *another*[c] as far as becoming in regard to substantial being is concerned, or that it comes to be *thus* from having been *another*[d] as far as becoming in regard to accidental being is concerned, since every change has two terms, it so happens that we can say this in two ways. The reason for this is that the terms of any becoming or change can be taken as simple or composite.[e,142]

---

[a]Less. 1, nn. 7 ff.

[b]"propositum."

[c]"aliud...ex alio."

[d]"alterum ex altero," i.e., something has one quality and comes to have another quality.

[e]Cf. St. Thomas, *Summa Theol.*, IIIa, qu. 17, art. 1, ans. 7.

And this is expressed in the following way: Sometimes we say that a *man becomes musical*, and then the two terms of the becoming are simple, and, likewise, when we say that *the non-musical becomes musical*. But when we say that *a non-musical man becomes a musical man*, then each of the terms is composite. When becoming is attributed to man or to non-musical, each term is simple. And thus, what comes to be, i.e., that to which becoming is attributed, is said to become as simple. But that in which the becoming itself is terminated, which is said to become as simple, is musical, as when I say *man becomes musical* or that *non-musical becomes musical*. But both are said to become as a composite (viz., *both what comes to be*, i.e., what becoming is attributed to, *and what came to be*, i.e., that at which the becoming is terminated), when we say that a *non-musical man becomes musical*, for then there is composition from the viewpoint of the subject only, and simplicity from the viewpoint of the predicate. But when I say that a *non-musical man becomes a musical man*, there is composition from the viewpoint of both.[a,143]

102(5). Then, when he says: (3), he shows two differences in what is said above. The first of these is that in some of the premises we use an ambiguous method of speaking, viz., *this become this*, and *this comes to be from this*, for we say that *non-musical becomes musical*, and *musical comes to be from non-musical*. We do not, however, speak in this way in all cases, for we do not say that *musical comes to be from man*, but rather that *man becomes musical*.

He states the second difference (4). And he says that when becoming is attributed to two simple things, viz., to a subject and its opposite, one of these

[a]Both of subject and of predicate.

remains and the other does not, because, when someone has already becom musical, the man persists, yet the opposite does not persist, whether it be a negati opposite, e.g., non-musical, or a privative or contrary opposite, e.g., unmusica Nor does the composite of the subject and its opposite remain, for a non-music man does not remain after the man has become musical. And yet, becoming attributed to these three things, for it was said that *man becomes musical*, and th *non-musical becomes musical*, and that *a non-musical man becomes musical*. O these three, only the first remains when the becoming is completed, but the othe two do not remain.

103(6). Then, when he says: (5), after the assumption of the premises h clarifies the point in question, viz., that there are three factors in any natur becoming whatsoever. And with regard to this, what he does is threefold. Firs he enumerates the two factors that are found in any natural becoming whatsoeve secondly, he proves what he had supposed (6), and thirdly, he concludes the poin in question (10).

104(7). First, he says, therefore, that, after the premises have been assumed if anyone wanted to consider all the things that come into being in the order o nature, he would admit this, viz., that there must always be some subject to whic the becoming is attributed, and that, although it is one in number or in subject, yet it is not the same in species or in meaning.[a] The reason for this is that when it i attributed to man that he becomes musical, man, indeed, is one in subject, but tw in meaning, for man and non-musical are not the same in meaning. He does not

---

[a]"ratione," i.e., in notion or definition.

however, state a third, viz., that in generation something must be generated, since that is obvious.

105(8). Then, when he says: (6), he proves two of the points which he had supposed: first, that the subject to which becoming is attributed has two meanings and secondly, that a subject must be supposed in any becoming (8). He makes the first clear in two ways: first, indeed, by the fact that in a subject to which becoming is attributed, there is something that remains and something that does not remain, because what is not opposed to the terminus of the becoming remains, for man remains when he becomes musical, but non-musical does not remain, nor does the composite, e.g., the non-musical man. And it is apparent from this that man and non-musical are not the same in meaning, since the one remains and the other does not.

106(9). Secondly, (7), he shows the same thing in another way, because in permanent things we say that *this comes to be from this* rather than *this becomes this* (although, we can say this, but not properly so), for we say that *musical comes to be from non-musical.*[144] We also say that the *non-musical becomes musical*, but this is accidental, viz., to the extent that what happens to be non-musical becomes musical. But we do not speak in this way in the case of permanent things, for we do not say that *musical comes to be from man*, but that the *man becomes musical.* Yet, in the case of permanent things, we sometimes say: *this comes to be from this* as *a statue comes into being from bronze*, but this occurs because by the name bronze we understand unshaped, and we say this in this way by reason of the understood privation.[a,145] And, although we say *this comes to be from this* in the case of permanent things, still, in non-permanent things, it so happens that we say both, viz., *this becomes this* and *this comes to be from this*, whether what does not remain is taken as the opposite or as the composite of the opposite and the subject. And, therefore, because we use this different way of speaking about a subject and the opposite, it is clear that it so happens that subject and opposite, e.g., man and non-musical, although they are the same in subject, still have two meanings.[146]

107(10). Then, when he says: (8), he elucidates the other supposition, i.e., that there must be a subject in every natural becoming, and, indeed, to prove this by argument is the task of metaphysics. And so, it is proved in *Metaphysics* VII,[b,147] but here he proves it by induction only. He first proves it from the viewpoint of what comes to be, and secondly, from the viewpoint of the modes of becoming (9). First he says, therefore, that since becoming may be said in many ways, becoming *strictly speaking*[c] is the becoming of substances only, but other things are said to become *in a limited sense*.[d] And the reason for this is that becoming implies the beginning of existence. For something to come to be absolutely, it is, therefore, required that the thing previously did not exist at all. This occurs in the case of what comes to be substantially.[e] For what becomes a

---

[a]Cf. St. Thomas, *In VII Metaph.*, less. 6, nn. 1414–1415.

[b]Less. 6, nn. 1381 ff.

[c]"absolute."

[d]"secundum quid."

[e]"substantialiter."

man not only was not a man previously, but it is entirely true to say that it did not exist. But when a man becomes white, it is not true to say that he did not exist previously, but that he was not thus previously. In the case of beings that come to be in a limited sense, it is, therefore, obvious that they require a subject, for quantity, quality and other accidents, of which there is becoming in a limited sense, cannot exist without a subject, for it is proper to substance alone not to exist in a subject. But even in the case of substances, if anyone would consider it, it becomes obvious that they come to be from a subject, for we see that plants and animals come to be from a seed.

108(11). Then, when he says: (9), he shows the same thing by induction from the modes of becoming. And, he says that of those things that come to be, some come to be by change of shape, as a statue is made from bronze. Some, however, come to be by addition, as is obvious in all things that are increased, as a river is made from many streams. And still other things come to be by taking away, as the image of Mercury comes to be from stone through the medium of sculpture. And some things come to be by composition, as a house. Still other things come to be by alteration, as those things of which the matter is altered, whether they come to be naturally or artificially. And in all these cases, it is apparent that they come to be from some subject. And so, it is clear that everything that comes to be comes to be from a subject. But we must note that he has enumerated artificial things along with those things that come to be absolutely (although artificial forms are accidents), either because artificial things are in some way in the genus of substance by reason of their matter, or he has done this because of the opinion of the ancients, who, likewise, considered natural things as artificial as will be said in the second book.[a]

109(12). Then, when he says: (10), he concludes the point in question. And he says that it is from what has been said that what is said to become is always a composite. And, since in any becoming, there is that at which the becoming terminates, and that which is said to become, which is twofold, viz., the subject and the opposite, it is clear that in any becoming, there are three factors, viz., *the subject*, *the terminus of the becoming*, and *its opposite*, e.g., when man becomes musical, the opposite is non-musical, and the subject is man, and musical is the terminus of the becoming. And likewise, the shapelessness, lack of form, and lack of order are the opposites, but bronze, gold, and stones are subjects in artificial things.

---

[a]Less. 2, n. 1.

TEXT OF ARISTOTLE

LESSON THIRTEEN[a]

1. If then, we grant that the things of Nature have ultimate determinants and principles which constitute them, and also that we can speak of them 'coming to be' not in an incidental but in an essential sense—so as to come to be the things they are and which their names imply, not having been so before—then it is obvious that they are composed, in every case, of the underlying subject and the 'form' which their defining properties give to it; for the cultivated man is in a way 'compact' of the subject 'man' and the qualification 'cultivated,' for the definition of such a *compositum* may always be resolved into the definitions of those two components. Clearly then these are the elements, or factors, out of which things that 'come to be' arise.

2. Now the subject is numerically one thing, but has two conceptually distinct aspects, for the man, or the gold, or the material factor in general is a thing that can be counted, since it may almost be regarded as a concrete individual thing and is not an incidental factor in the generation of what comes into being; whereas the negation of the emergent qualification or the presence of its opposite is incidental. On the other hand, the form—e.g. the 'order' or the 'culture' or any other such predicable qualification—is also one thing.

3. So there is a sense in which the ultimate principles of the sum of changing things are two, but a sense in which they are three; for the actual change itself takes place between the terms of an antithesis, such as cultivated and uncultivated, hot and cold, articulated and unarticulated, and so forth; but from another point of view these two principles are inadequate, for they cannot possibly act or be acted upon directly by each other. This difficulty, however, disappears if we admit, as a third principle, a non-antithetical 'subject.' So in a sense there are no principles except the terms of opposition, and it may be said that they are two in number and no more; but there is a sense also in which we cannot quite admit this and must go on to three, because of the conceptual distinction that exists in them; for instance in the

---

[a]Includes 11. 190 b 17—191 a 22, Wicksteed and Cornford, *op. cit.*, pp. 79-83.

uncultivated man, between his being a man and his being uncultivated, or in the unshaped bronze, between its being shaped and its being bronze.

4. It is now clear, then, how many are the principles of things in the changing world of Nature, and in what sense; namely that there is something that underlies all opposites, and that opposition involves two terms.

But if we take it another way, we may escape the duality of the opposition by considering one of its terms taken singly as competent, by its absence or presence, to accomplish the whole change.

5. Then there will only be the 'ultimately underlying' factor in Nature in addition to this formal principle to reckon with. And of this 'underlying' factor we can form a conception by analogy; for it will bear the same relation to concrete things in general, or to any specific concrete thing, which the bronze bears to the statue before it has been founded, or the wood to the couch, or the crude material of any object that has determined form and quality to that object itself. This ultimate material will count as one principle (not, of course, one in the sense of a concrete 'individual'); and the collectivity of determining qualities implied by the thing's definition is also one principle; and further there is the opposite of this, namely the 'being without' or 'shortage' of it.

How the principles, then, can be taken as two, and how that enumeration appears to need supplementing, has now been shown.

6. First it appeared as though the 'terms of an antithesis' constituted all the principles necessary; but then we saw that something must underlie them, constituting a third. And now we see that the two terms of the opposition itself stand on a different footing from each other; and we see how all the principles are related to each other, and what we are to understand by the 'underlying subject.' It remains to consider whether the modifying 'form' or the modified 'matter' is to be regarded as three principles altogether, and in what sense they are three, has already been demonstrated.

Let this, then, suffice as to the number of the principles and as to what they are.

In the being and becoming of natural things there are two substantial principles: matter and form, and one accidental principle: privation.

110(1). After the Philosopher has shown that there are three factors in any natural becoming, *it is his intention to make clear from the preceding the number of the principles of nature.*[148] And with regard to this, what he does is twofold. First, he makes clear the point in question, and secondly, by recapitulating, he shows what has been said, and what remains to be said (6). And again with regard to the first, he has a twofold procedure. First, he shows that there are three principles of nature, and secondly, he elucidates these (5). He deals with the first in three ways. First, he shows what the truth is with regard to the principles of nature; secondly, from this truth, he resolves the premised questions with regard to the principles (3); thirdly, because the ancients said that principles are contraries, he shows whether or not contraries are always required (4).

And again, with regard to the first of these, i.e., showing what the truth is in reference to principles of nature, what he does is twofold. First, he shows that the substantial[a] principles of nature are two, and secondly, he shows that there is a third accidental principle of nature (2).

---

[a]"per se."

111(2). For the first, he uses the following argument: Those are said to be principles and causes of natural things from which they are and come to be substantially and not accidentally.

But everything that comes to be is and comes to be from a subject and form. Subject and form are, therefore, substantial causes and principles of everything that comes to be in nature. But that what comes to be in the order of nature comes to be from subject and form, he proves in the following way: Those things into which the definition of something is resolved compose that thing, because any one thing is resolved into those things of which it is composed. But the essence of what comes into being in nature[a] is resolved into subject and form,[149] for the essence of a musical man is resolved into the essence of man and into that of musical, for, if anyone wanted to define musical man, he would have to give the definition of man and of musical. What comes to be in nature is, therefore, and comes to be from subject and form. And we must note that here he intends to investigate the principles not only of becoming but also of being. And so, it is significant that he says *they are and come to be from these first things*. And he says *from these first things*, i.e., substantially and not accidentally. The substantial principles of everything that comes to be in nature are, therefore, subject and form.

112(3). Then, when he says: (3), he adds a third, an accidental principle.[b,150] And he says that, although the subject is one in number, yet, in species

---

[a]"secundum naturam."

[b]Cf. St. Thomas, *De Principiis Naturae*, chap. 2, critical text of John J. Pauson (Fribourg: Société philosophique, 1950), pp. 82–85.

and in meaning, it is twofold, as was said above,[a] because man and gold and any material thing have a certain numerical predication. Indeed, there is the subject itself to consider. This is something positive. From it something comes to be substantially and not accidentally, e.g., man and gold. And there is also its accident to consider, viz., the contrariety and the privation, e.g., unmusical and unmoulded. And the third is the species or the form, as order is the form of a house, or music of a musical man, or anything else that is predicated in this manner.

In this way, then, *form* and *subject* are *substantial* principles of what comes to be in nature, but the *privation* or the *contrary* is an *accidental* principle to the extent that it is an accident of the subject, e.g., we say that a builder is substantially the active cause of a house, but musical is accidentally the active cause of the house to the extent that to be musical is accidental to the builder. And so, as a subject, man is a substantial cause of the musical man, but non-musical is his accidental cause and principle.

113(4). One might object, however, that privation is not an accident to a subject when it has a form and that for this reason privation is not an accidental principle *of being*.[b,151] It must be said, therefore, that matter is never without privation, because when it has one form it has the privation of another form. And, therefore, while something which has come to be is in the process of becoming (e.g., a musical man), and when that form is not yet in the subject, there is the privation of music itself. And so, the accidental principle of a musical man in the

---

[a]Less. 12, n. 7.

[b]"essendi."

process of becoming is non-musical, for this is an accident to man while he becomes musical. But when this form has come to him, the privation of another form is added, and so, the privation of the opposite form is the accidental principle in being.[a] It is clear, therefore, according to the intention of Aristotle, that privation, which is posited as an accidental principle of nature, is not some aptitude to form or beginning of form or some imperfect active principle, as some say, but the very lack of form or the contrary of form, which is an accident of the subject.

114(5). Then, when he says: (3), he resolves all the preceding doubts with the truth that he has established. And he concludes from what is said above, that we must say that in one way there are two principles, viz., the substantial ones, and in another way three, if an accidental principle is assumed along with the substantial principles. And in one way, the principles are contraries, e.g., if we would take the musical and the non-musical, the hot and the cold, the harmonious and the discordant. And in another way, the principles are not contraries, i.e., if they are taken without a subject, because contraries cannot be mutually acted upon, unless it is explained by the fact that underlying the contraries there is some subject, by reason of which they are mutually acted upon.

And thus, he concludes that there are no more principles than there are *contraries*;[b] but there are only two substantial principles. But they are not two in every sense, because one of them according to its being[c] is the other, for the subject

---

[a]"essendo."

[b]"principia non sunt plura contrariorum, idest contrariis, hoc est quam contraria."

[c]"secundum esse."

is in meaning[a] twofold, as was said.[b] And so, there are three principles, because man and non-musical, and bronze and unshaped differ in meaning.[152] It is clear, therefore, that the previous expositions, which argued in behalf of each viewpoint, were partly true but not completely.

115(6). Then, when he says: (4), he shows in what way contraries are necessary, and in what way they are not. And he says that from his previous remarks,[c] it is clear how many principles of generation of natural beings there are, and in what way they are so many. For it has been shown[d] that two contraries are necessary, one of which is a substantial principle and the other an accidental one, and that something underlies[e] the contraries. And this is also a substantial principle. But in a way, one of the contraries is not necessary for generation, for under certain circumstances the other of the contraries by its absence or presence is sufficient to effect a change.

116(7). To clarify this, we should know, as is said in the fifth book of this work,[f] that there are three kinds of change, viz., *generation*, *corruption* and *motion*. The difference in these is that *motion* is from one affirmative to another

---

[a]"secundum rationem."

[b]Less. 12, n. 7; less. 13, n. 3.

[c]This less., nn. 2–5.

[d]He is referring to less. 10 and 11 and, in fact, to all that precedes less. 12.

[e]"subjiciat."

[f]Less. 2, nn. 2–3.

affirmative, e..g, from white to black. *Generation*, however, is from a negative to an affirmative, e.g., from non-white to white, or from non-man to man. But *corruption* is from an affirmative to a negative, as from white to non-white, or from man to non-man. And so, it is clear that two contraries and one subject are required in motion. But in generation and corruption the presence of one contrary and its absence, which is a privation, are required. Moreover, generation and corruption are preserved in motion, for from the fact that there is movement from white to black, the white is corrupted and the black comes to be. So, therefore, in every natural change, subject, form and privation are required. But the meaning of motion is not preserved in every generation and corruption, as is evident in the generation and corruption of substances.[153] And so, subject, form and privation are preserved in every change but not the subject and its two contraries.

117(8). This opposition is also found in substances, substance being the first genus, but not the opposition of contrariety,[a] for substantial forms are not contraries, although differences in the genus of substance are contraries, in that one is taken with the privation of the other, as is evident in the case of the animate and the inanimate.

118(9). Then, when he says: (5), he elucidates the foregoing principles. And he says that the nature which is primarily subjected to change, i.e., prime matter, cannot be known in itself, since everything that is known is known by its

---

[a]Cf. St. Thomas, *In X Metaph.*, less. 10.

form. And prime matter is considered to be the subject of every form.[a,154] But it is known *analogously*,[155] i.e., by proportion.[b] For it is in this way that we know that wood is something other than the form of bench or couch, because sometimes it is under one form and sometimes under another. When we see that air sometimes becomes water, we must, therefore, say that something existing under the form of air sometimes exists under the form of water. And thus, we must also say that there is something besides the form of water and besides the form of air, just as wood is something other than the form of bench and other than the form of couch. Whatever is so related to those natural substances as bronze is related to a statue and wood to a couch, and anything at all material and unformed to form, we say, therefore, that this is *prime matter*.

This is, therefore, one principle of nature. It is not one *as this something*,[c] i.e., as some indicated individual, in such a way that it has form and unity in act, but it is called being and one to the extent that it is in potency to form. But the

---

[a]Cf. St. Thomas, *De Princ. Nat.*, *op. cit.*, pp. 85–86; *In VII Metaph.*, less. 2, nn. 1277 and 1276 and read them in this order.

[b]On the question of analogy, cf.: Cajetan, *De Analogia Nominum*, chap. 1–5; John of St. Thomas, *Curs. Phil.*, Vol. I, Log. 2, qu. 13, art. 3–4; *ibid.*, qu. 14, art. 2–3; St. Thomas, *In IV Metaph.*, nn. 534 ff.; *Summa Theol.*, Ia IIae, qu. 85; *In I Ethic.*, less. 6; *ibid.*, less. 7, nn. 95–97; *In V Ethic.*, less. 5; *In I Sent.*, dist. 19, qu. 5, art. 1–2; *De Ver.*, qu. 1, art. 2, 4; *ibid.*, qu. 3, art. 5; *ibid.*, qu. 2, art. 11; *ibid.*, qu. 9, art. 4–5; *ibid.*, qu. 21, art. 4, ans. 2; *ibid.*, qu. 27, art. 2, 5; *Cont. Gent.*, Bk. I, chap. 34; *De Trin.*, qu. 5, art. 3; L. Geiger, O.P., *La Participation dans la philosophie de St. Thomas d'Aquin* (Paris: J. Vrin, 1953); W. Norris Clarke, S.J., "The Meaning of Participation in St. Thomas," *Proceedings of the American Catholic Philosophical Association*, Vol. XXVI (1952), pp. 147–157; Normand Marcotte, *op. cit.*

[c]"hoc aliquid."

nature[a] or form is another principle, and a third is privation, which is contrary t the form. And in what sense these principles are two and in what sense they ar three has been said previously.[b]

119(10). Then, when he says: (6), he sums up what has been said, an shows what remains to be said. And so, he says that it has been stated above[c] tha contraries are principles and then it was said that they have some subject.[d] And so there are three principles. And from what has now been said,[e] the differenc between the contraries is clear: one is a substantial principle and the other a accidental one. And also, we have stated how principles are related to one another the subject and contrary are one in number and two in meaning. And also, wha the subject is has been stated as clearly as possible. But we have not yet sai whether form or matter is more substance, for this will be said in the beginning o the second book.[f] And so, we have said that there are three principles and in wha sense, and in what way each is a principle.[g] And lastly, he concludes his principa

---

[a]"ratio."

[b]Less. 12, n. 7; this less., n. 3 and end of n. 5.

[c]Less 12, nn. 6—end of less.

[d]"quod eis aliquod subjicitur."

[e]This less., nn. 2–9.

[f]Less. 2, n. 5.

[g]Cf. Vincent Smith, *Philosophical Physics* (New York: Harper and Bros. 1950), pp. 47–80; Kenneth Dougherty, *Cosmology* (Peekskill, N.Y.: Graymoo Press, 1952), pp. 102–134; St. Thomas, *De Princ. Nat.*, the entire work.

intention, viz., that it is clear how many principles there are and what they are.[a]

[a]Cf. H. D. Gardeil, O.P., *Introduction to the Philosophy of St. Thomas Aquinas*, Vol. II, *Cosmology*, translated by John A. Otto (St. Louis: B. Herder Book Co., 1948), pp. 17–42; W. D. Ross, *Aristotle's Physics* (Oxford: Clarendon Press, 1936), pp. 19–26, 337–348; Peter Hoenen, S.J., *The Philosophical Nature of Physical Bodies* (West Baden Springs, Ind.: West Baden College, 1955).

EXT OF ARISTOTLE LESSON FOURTEEN[a]

1. It remains to show that the conclusion we have reached not only solves e problem of genesis, but furnishes the only escape from the blind alley into hich the first speculations on the subject led their authors. For when first they egan to reason on the truth of things and the nature of all that exists, pioneers as ey were, they fell upon a false track for want of a clue, and maintained that othing at all could either come into existence or pass out of it; for they argued at, if a thing comes into existence, it must proceed either out of the existent or ut of the non-existent, both of which were impossible; for how could anything ome out of' the existent, since it is already there? and obviously it could not come ut of the non-existent, for what it comes out of must be there for it to come out f, and the non-existent is not there at all. And so, developing the logical onsequences of this, they went on to say that the actually and veritably 'existent' not many, but only one.

2. Such then is their dogma; but, as for us, we maintain that when we speak f anything 'coming to be,' whether out of existence or non-existence, or of the on-existent or the existent acting or being acted on in any way, or of anything at ll 'becoming this or that,' one explanation is as follows:

It is much the same as saying that a 'physician' does or experiences omething, or that he has 'become' (and now is) something that he has 'turned ito,' instead of remaining a physician. For all these expressions are ambiguous, nd this ambiguity is clearly analogous to the ambiguous, and this ambiguity is learly analogous to the ambiguity concealed under our language when we speak f what 'the existent has turned into,' or of the existent 'doing' this or 'experienc-g' that. For if the physician builds a house, it is not *qua* physician but *qua* uilder that he does so; or if he becomes light in complexion, it is *qua* dark in omplexion not *qua* physician that he changes; whereas if he exercises the healing rt, or drops or loses that art, so as to become a non-physician, it is *qua* physician at he does so. And so, just as strange conceptions might be formed as to what

---

[a]Includes 11. 191 A 23—191 b 34, Wicksteed and Cornford, *op. cit.*, p. 85–89.

a physician could or could not do or suffer or become, if we were always thinkin of him in his primary and direct capacity as a physician, but applied our concl sions to him in *all* his actual or possible capacities, so, obviously, if we alway argue from the non-existent *qua* non-existent, but apply our conclusions to th incidentally non-existent as well, we shall fall into analogous errors. And it wa just because the earlier thinkers failed to grasp this analytical distinction, that the piled misconception upon misconception to the pitch of actually concluding th there was no such thing as genesis and that nothing at all ever can to be, or wa except the one and only 'existent.'

Now we too (who recognize both 'form' and 'lack of form,' or 'shortage as factors in becoming), assert that nothing can 'come to be,' in the absolute sens out of the non-existent, but we declare nevertheless that all things which come t be owe their existence to the incidental non-existence of something; for they ow it to the 'shortage' from which they started 'being no longer there.' And if it seem an amazing paradox to maintain that anything derives in this way from the no existent, yet it is really quite true.

Moreover, it is equally true that it is only in this same incidental sense th anything can derive from the existent either, or 'what is' can come into being. I this sense, however, this does occur in the same way as (for instance) if 'an anima should turn into 'an animal,' or a particular animal—say, a horse—should turn int another particular animal—say, a dog. The dog would come into being, not onl 'out of' a particular animal, but out of 'an animal' (and it would become 'a animal'), but only incidentally, not *qua* animal, since it was already an animal an could not 'turn into' what it already was. If anything, then, is to 'turn into' a animal, otherwise than incidentally, it must be non-animal at the start and mu come to be an animal in the process. Similarly, if a thing is to become or turn int an existent otherwise than incidentally, it cannot start from what exists;—thoug neither can it start from the non-existent, for we have explained that this mean 'from the non-existent as such.' At the same time we do not do away with th principle that everything must either be or not be.

This then is one way of formulating the solution to the problem.

3. But there is also an alternative formula based on the distinction betwee existing as a potentiality and existing as an actuality. But this is developed mor fully elsewhere.

4. It is thus (as I have said) that the difficulties are solved which led to th denial of some of the obvious facts that we have now discussed; for it was thi

fundamental misconception that threw the earlier thinkers so far off the track concerning 'coming into existence' and 'passing out of it,' and the nature of change in general. Had they detected the existence of the entity we have described, it would have released them from all confusion.

[illegible] doctrine [illegible] the [illegible] ... and [illegible] [illegible] in [illegible] [illegible] the existence of the [illegible] would have [illegible] from [illegible] conclusion.

From the truth established with regard to principles, he resolves the doubts and errors of the ancients arising from their ignorance of matter.

120(1). After the Philosopher has established the truth with regard to the principles of nature, *here he excludes the difficulties of the ancient philosophers* through what was determined with regard to the principles.[a,156] He first excludes the difficulties or errors which have arisen from their ignorance of matter, and secondly, those which have arisen from their ignorance of privation (less. 15, n. 1); thirdly, he reserves for another science the questions which occur with regard to form (less. 15, n. 7).[157]

He deals with the first, viz., the treatment of the problems arising from the ignorance of matter, in two ways. First, he states the difficulty and error into which the ancient philosophers fell because of their ignorance of matter, and secondly, he resolves their problem by the truth established (2).

121(2). First, he says, therefore, that it is only through the truth established with regard to principles that every *error*,[b] i.e., every difficulty of the ancient philosophers is resolved. And this indicates that what has been said about

---

[a]Cf. St. Thomas, *In III Metaph.*, less. 1, n. 338.

[b]"defectus."

principles is true, for truth excludes every falsity and doubt. On the other hand, if anything false is stated, some difficulty must remain.

Now, the difficulty and error of the ancient philosophers was this. Those who first philosophically investigated the truth and nature of things turned their attention to another path away from the path of truth and the natural path. The reason for this was their poor understanding, since they said that nothing was either generated or corrupted. This is both contrary to truth and contrary to nature.

And their poor understanding compelled them to state this, since they did not know how to resolve the argument that seemed to prove that being was not generated. For, if being comes to be, it comes to be either from being or from non-being.[158] And both of these seem to be impossible, viz., that being comes to be from being, and that being comes to be from non-being. For that it is impossible that something come to be from being is clear from the fact that what is does not come to be, since nothing is before it comes to be. And being already exists, and so, it does not come to be. That it is also impossible that something come to be from non-being is clear from the fact that there must always be some subject for what comes to be, as was shown above,[a] and nothing comes to be from nothing. And they concluded from this that there was neither generation nor corruption of being.

And arguing further, they even went on to say that there were not many beings but only one being. And they said this on account of the foregoing. For, since they said that the material principle was one and that from it nothing is caused after the manner of generation and corruption but only after the manner of alteration, it followed that it would always remain one according to substance.

---

[a]Less. 12, nn. 4, 10.

122(3). Then, when he says: (2), he resolves the foregoing objection. And with regard to this he proceeds in a twofold manner. First, he resolves that objection in two ways, and secondly, he concludes his main proposition (4). The first way is divided into two parts, according to the two solutions which he states; the second he states at (3).

123(4). First, he says, therefore, that as far as the manner of speaking is concerned, there is no difference in saying that something comes to be from being or from non-being,[159] or that being or non-being act[a] or are acted upon, or in saying the same thing of anything else,[b] and in stating propositions of this kind about a physician, viz., that a physician acts or is acted upon or that from the fact that he is a doctor he is or becomes something.

But to say that a physician acts or is acted upon or that he becomes something from being a physician has a twofold meaning. To say that something comes to be from being or from non-being, or to say that being or non-being act or are acted upon, has, therefore, a twofold meaning. And, likewise, it may be stated in other terms, e.g., if we say that something comes to be from white, or that white acts or is acted upon. And so, in this way it is clear that there is a twofold meaning in saying that a physician acts or is acted upon, or that he becomes something from being a physician. For we say that a physician builds, but he does not do this to the extent that he is a physician but in that he is a builder, and likewise, we say that a physician becomes white, not to the extent that he is a physician but in that he is black. In another way, we say that a physician heals in

---

[a]"faciat aliquid."

[b]By substituting other terms for "being" and "non-being."

that he is a physician, and likewise, that a physician becomes a non-physician to the extent that he is a physician. But strictly speaking,[a] we say that the physician acts or is acted upon or that something comes to be through the physician, when this is attributed to him as physician. But, when this is not attributed to him as a physician but as something else, we say that this is attributed to him in an incidental manner. So, it is clear, therefore, that when it is said that a physician acts or is acted upon, or that he becomes something from the fact that he is a physician, this is understood in two ways, viz., in the full sense or incidentally.

And so, it is obvious that when it is said that something comes to be from non-being, this is understood in the strict sense as if something comes to be from non-being as such. And the same reasoning is true with regard to being.[160]

And the ancient philosophers, not perceiving this distinction, were in error to such an extent that they thought that nothing came to be. Nor did they think that anything else, besides what they posited as the first material principle, had substantial being. For example, in saying that air was the first material principle, they said that all other things signify a kind of accidental being, and thus they excluded all substantial generation, leaving alteration only. From this fact, viz., that something does not come to be in the full sense either from being or from non-being,[b]

---

[a]"proprie et per se."

[b]Cf. the reasoning of Sextus in the summary of frag. 3: "(Sextus, from *On Being* or *On Nature*): I. Nothing exists. (a) Not-Being does not exist (b) Being does not exist. i. as everlasting. ii. as created. iii. as both. iv. as One. v. as Many. (c) A mixture of Being and Not-Being does not exist. II. If anything exists, it is incomprehensible. III. If it is comprehensible, it is incommunicable." Freeman, *op cit.*, p. 128. Cf. also what Isocrates says of Gorgias: frag. 1: "(Isocrates: Gorgias had the hardihood to say that nothing whatever exists)."

they thought that nothing could come to be from being or non-being.[a,161]

124(5). But we say that from non-being nothings come to be absolutely and in the full sense but only incidentally, because what is, i.e., being, is not in the full sense of privation. And this is so because privation does not enter into the essence of the thing that has come to be,[b] but something comes to be in the full sense from what is in the thing after it has already come to be. For example, shaped comes to be from unshaped, not in the full sense but incidentally, because after it is already shaped, the unshaped is no longer in it. But that is the extraordinary way of becoming something from non-being, and it seemed impossible to the ancient philosophers. And so, it is clear that from non-being something comes to be not substantially but accidentally.

125(6). Likewise, if we investigate whether or not something comes to be from being, we must say that from being something comes to be incidentally, but not in the full sense. And he makes this clear by the following example: For, if we suppose that a dog is generated from a horse, it is obvious that a certain animal

---

"(Isocrates: Gorgias had the hardihood to say that nothing whatever exists)." "Gorgias of Leontini: latter half of fifth century B.C. He wrote one of the earliest Handbooks on Rhetoric; an essay *On Being* or *On Nature*; and a number of model orations, of which parts have survived: from the Olympian Oration, the *Encomium on Helen*, and the *Defence of Palamêdês*." *Ibid.*, p. 127.

[a]Cf. St. Thomas, *In XII Metaph.*, less. 3, n. 2443.

[b]Cf. less. 12.

comes to be from a certain animal. And in this way, an animal will come to be from an animal. Yet, an animal will not come to be in the full sense from an animal, but only incidentally, for it does not come to be to the extent that it is animal but to the extent that it is this animal, because it is already animal before it becomes a dog, since it is a horse, although it is not yet this animal, i.e., dog. And so, this animal that is a dog comes to be in the full sense from what is not this animal, i.e., non-dog. But if an animal came to be in the full sense and not incidentally, it would have to come to be from non-animal.

The same is true with regard to being, for some being comes to be from what is not this being. But it is incidental to what is not this[a] that it becomes a being. Thus, something does not become in the full sense from being, nor in the full sense from non-being, for this becoming in the full sense signifies that something comes to be from non-being, if it comes to be from non-being as such, as was said.[b] And, just as when this animal comes to be from this animal, or this body from this body, neither every body nor every non-body, nor every animal or non-animal is discarded from that from which something comes to be, so also neither all existence nor all non-existence is discarded that from which this being which is fire comes to be has some being, since it is air, and some non-being, since it is not fire.[162]

126(7). This, therefore, is one way of resolving the foregoing doubt. But this method of resolving it is insufficient, for, if being comes to be incidentally both from being and from non-being, we must posit something from which being comes

---

[a]What is not this being that has come to be.

[b]This less., n. 4.

to be in the full sense, because everything that is incidentally is reduced to what is in the full sense.

127(8). In order to show what it is from which something comes to be in the full sense, he adds a second mode (3). And he says that the same things can be spoken of both from the point of view of potency and from the point of view of act, as is determined in a more definite way elsewhere, viz., in the *Metaphysics*, IX.[a,163] Something comes to be in the full sense, therefore, from being in potency, but something comes to be incidentally from being in act or from non-being. And he says this, because matter, which is being in potency, is that from which something comes to be in the full sense,[b,164] for this is what enters into the substance of the thing made. But something comes to be incidentally from privation or from a preceding form to the extent that it was the form of matter, from which something comes to be in the full sense, to have such a form or such a privation, e.g., a statue comes to be in the full sense from bronze, but a statue comes to be incidentally from not having such a shape and from having another shape.

128(9). Finally, he concludes his main proposition (4), when he states that insofar as what we are saying is true, then all problems, i.e., *difficulties*, are resolved because of the above statements. Compelled by these doubts, some ancient philosophers cast aside some of the foregoing notions, viz., generation and

---

[a]The entire book, but esp. less. 1.

[b]Cf. St. Thomas, *In XII Metaph.*, less. 2, n. 2437; *In XI Metaph.*, less. 6, n. 2227.

corruption, and the plurality of things substantially different. But now that this nature, i.e., *matter*,[a] is made clear, he resolves all their ignorance.

[a]Cf. R. P. Phillips, *Modern Thomistic Philosophy*, Vol. I, *The Philosophy of Nature* (Westminster, Md.: The Newman Press, 1950), pp. 36–53.

TEXT OF ARISTOTLE — LESSON FIFTEEN[a]

1. Certain other thinkers have approached the view we have now expounded, but without effectively reaching it. For they began by accepting as true the conception of Parmenides that 'coming to be' could mean nothing but emerging from the non-existent. Then, under the one concept of the 'non-existent,' they united both the 'matter' and the 'shortage' of our system, failing to distinguish between 'numerical unity' of subject (which we admit) and 'unanalysable unity' of aspect or potentiality (which we deny).

2. But this distinction which they ignored makes all the difference. For we distinguish between 'matter' and 'shortage' (or absence of form), and assert that the one, namely matter as such, represents the incidental non-existence of attributes, whereas the other, namely shortage as such, is the direct negation or non-existence though never existing in isolation, may be pretty well taken as constituting the 'concrete being' of which it is the basis, but shortage not in the least so.

3. These philosophers, on the other hand, conceive of the non-existent as the 'great' and the 'small' alike, whether severally or conjointly. So they too have a triad of the 'great,' the 'small,' and the 'Idea' (or form), but this triad is really quite different from ours of 'matter,' 'shortage,' and 'form'; for, although they go so far with us as to recognize the necessity of some underlying subject, yet in truth the 'great and small' of which it consists can only be equated with our 'matter,' and is not a dyad at all. It is true indeed that you may now and again find it spoken of as the 'dyad' of great and small, but that makes no difference, for the other member of our antithesis (shortage, namely) is systematically ignored.

4. Now we, who distinguish between matter and shortage, can very well see why matter, which cooperates with form in the genesis of things, may be conceived as their matrix or womb. And we can also see how a man who concentrates his mind on the negative and defect-involving character of shortage may come to think

---

[a]Includes 11. 191 b 35—192 b 7, Wicksteed and Cornford, *op. cit.*, pp. 91–97.

of it as purely non-existent. Then if we were to think of 'existence' as something august and good and desirable, we might think of shortage as the evil *contradiction* of this good, but of matter as a something the nature of which is to *desire* and yearn toward the actually existent.

5. But the school of thought we are examining, in as much as it identifies matter and shortage falls into the position of representing the opposite of existence as yearning for its own destruction. But how can either form or shortage really desire form? Not form itself, because it has no lack of it; and not shortage, which is the antithesis of form, because the terms of an antithesis, being mutually destructive, cannot desire each other. So that if (to borrow their own metaphors) we are to regard matter as the female desiring the male or the foul desiring the fair, the desire must be attributed not to the foulness itself, as such, but to a subject that is foul or female incidentally.

6. As for the undifferentiated matter-shortage of the school we have been discussing, it may be regarded as either perishable or imperishable; for if we think of it as the bare *seat-of-shortage*, it perishes, as such, for 'shortage' is exactly what does perish in it, on its receiving the form; but if we are considering it as the potentiality of receiving forms, it cannot perish, as such, but must necessarily be exempt both from destruction and genesis. For if it 'came to be,' there must have been some subject already there for it to proceed from, and just 'being there as the subject' is precisely what constitutes the nature of 'matter' itself, so that it must have been before it came to be. For what I mean by matter is precisely the ultimate underlying subject, common to all things of Nature, presupposed as their substantive, not incidental, constituent. And again, the destruction of a thing means the disappearance of everything that constitutes it except just that very underlying subject which its existence presupposes, and if this perished, then the thing that presupposes it would have perished with it by anticipation before it came into existence.

7. So much for 'matter'; but the detailed examination and determination of 'form' as a principle, and the question of its unity or plurality, and its nature (singular or plural), is the business of First Philosophy; so let it be deferred till we come to that. It is with natural and perishable forms only that we shall deal in the sequel of this treatise.

Let this suffice for the demonstration of the existence of principles in Nature and the determination of what they are and how many in number. In the next book we must make a fresh start with fresh questions.

Matter is distinguished from privation, and it cannot be generated or corrupted substantially.

129(1). After the Philosopher excludes the doubts and errors of the ancient philosophers resulting from their ignorance of matter, *here he excludes the errors resulting from their ignorance of privation.* And he deals with this in three ways. First, he sets forth the errors of those who were wrong; secondly, he shows the difference between this position and the truth, which he established above (2); thirdly, he proves that his opinion is the correct one (4).

130(2). First, he says, therefore, that some philosophers touched upon the subject of matter, but did not treat it sufficiently, because they did not distinguish between privation and matter. As a result, what is proper to privation they attributed to matter. And, because privation in itself is non-being, they said that matter in itself was non-being. And so, just as something comes to be absolutely and in the full sense[165] from matter, so they believed that something comes to be absolutely and in the full sense from non-being. And they were led to believe this by two arguments. They were first led to believe this by the argument of Parmenides, who said that whatever is other than being is non-being. And so, since matter is other than being, because it is not being in act, they said that it was absolutely non-being. The second argument that led them to believe this was that what is one in number or in subject seemed to them to be one in meaning.[166] This latter he calls being *potentially* one, because things that are one in meaning are such that the power[a] of each is the same. But things that are one in subject and not in meaning do not have the same potency or power, as is seen in the case of white and of musical. But a subject and its privation, e.g., bronze and unshaped, are one numerically. Thus, it seemed to them that these were the same in meaning or in power. It is in this way that he understands the unity of potency.

131(3). But lest someone, because of these words, might doubt what the potency of matter is, and whether it is one or many, we must say that act and potency divide any genus whatsoever of beings, as is evident in *Metaphysics* IX,[b,167] and in the third book of this work.[c,168] And so, just as the potency for quality is not something outside the genus of quality, so the potency to substantial being is not something outside the genus of substance. The potency of matter is, therefore, not a property added to its essence, but the essence of matter is potency to substantial being. And yet, the potency of matter in subject is one in respect to many forms, while in meaning, there are many potencies according to the disposition of different forms.[169] And so, in the third book of this work,[d,170] it will be said that the capacity to be healed and the capacity to become ill are different in meaning.

---

[a]"virtus."

[b]Less. 1, n. 1770.

[c]Less. 1, nn. 279–280; *ibid.*, less. 2, n. 285; *ibid.*, less. 3, n. 296.

[d]Less. 2, n. 290.

132(4). Then, when he says: (2), he shows the difference between his opinion and the preceding one. And he deals with this in two ways. First, he explains his own opinion, and secondly, he shows what the other opinion states (3). So, in the first place, he says that it makes a great deal of difference whether something is one in number or in subject, and whether it is one in potency or one in meaning,[171] since we ourselves say, as is evident from the preceding,[a] that matter and privation, although they are one in subject, are, nevertheless, different in meaning. This is obvious for two reasons. The first reason is that matter is non-being incidentally, but privation is non-being in the full sense. For example, what is unshaped signifies non-being, but bronze does not signify non-being, unless inasmuch as it happens to be unshaped. The second reason, indeed, is that matter is *nearly a thing*,[b] and is[c] in some way, because it is in potency to be a thing and is, in a way, the substance of a thing, because it enters into the constitution of the substance. But this cannot be said of privation.[172]

133(5). Then, when he says: (3), he makes clear the meaning of the Platonic opinion. And he says that some followers of Plato posited a dyad[d] from the point of view of matter, viz., great and small,[173] but they posited them in a way different from that of Aristotle. For Aristotle stated these two to be matter and privation, which are one in subject, but different in meaning. The followers of

---

[a]Less. 12, n. 4; less. 13, n. 3 and end of n. 5.

[b]"prope rem."

[c]"Being" can in a certain sense be said of it.

[d]"duo," i.e., a duality.

Plato, on the other hand, did not state that one of these was privation and that th other was matter, but they took privation for both, viz., for small and great whether they took these two simultaneously, in that they did not distinguis privation by great and small, or whether they took each separately. And so, it i clear that the Platonists, who posited form, great and small, and Aristotle, too, wh stated matter, privation and form, posited three principles in an entirely differen way.[174]

But the followers of Plato arrived at a knowledge beyond that of the othe ancient philosophers in that they said that some one nature must be supposed to al natural forms. And this nature is prime matter. But they make this one only, bot in subject and in meaning, and do not distinguish between it and privation. Eve though they posit a duality from the viewpoint of matter, i.e., great and small, yet they make no distinction between matter and privation, but they only mentio matter, in which the great and the small are included, and they overlooked privatio and did not mention it.

134(6). Then, when he says: (4), he proves that his view is correct. An with regard to this, he follows his usual twofold procedure. First, he proves wha he has proposed, i.e., that privation must be distinguished from matter, an secondly, he shows how matter is corrupted or generated (6). And the first poin he proves in two ways: first, indeed, ostensively,[a] and secondly, by leading to th impossible (5).

---

[a]In a positive way.

135(7). And so, Aristotle says in the first place, that this nature, which is as subject, i.e., matter, is, together with form, the cause of what comes to be in the natural order in the way that a mother is a cause, for, just as a mother is the cause of generation in receiving, so is matter.

But if anyone should take the *other member of the contrariety*, i.e., privation, and concentrate his attention on this, he will imagine that privation does not belong to the constitution of a thing, but rather that it is a sort of evil. The reason for this is that privation is wholly non-being, since privation is nothing else but a negation of form in a subject, and is outside the whole being. Hence, in reference to privation, the argument of Parmenides holds, viz., that whatever is other than being is non-being; but it does not hold in reference to matter as the followers of Plato said.

And Plato shows that privation would then pertain to evil by the fact that form is something divine, best and desirable. It is divine, indeed, because every form is a certain participation in a similitude of divine being, which is pure act,[a] since a thing is in act to the extent that it has form.[b,175] And it is best, because act is the perfection of potency and is its good and, consequently, it follows that it is desirable, since a thing desires its perfection.[c,176] Privation, however, is opposed

---

[a]This term is, of course, that of St. Thomas.

[b]Cf. Marianne Miller, "The Problem of Action in the Commentary of St. Thomas Aquinas on the Physics of Aristotle," *The Modern Schoolman*, Vol. XXIII, No. 4 (May, 1946), pp. 224–226; cf. also Pierre Rousselot, S.J., *The Intellectualism of St. Thomas* (New York: Sheed and Ward, 1935), pp. 37–42; David Knowles, *The Historical Content of the Philosophical Works of St. Thomas Aquinas* (London: Blackfriars, 1958), pp. 8–9.

[c]Cf. John Warren, "Nature—A Purposive Agent," *The New Scholasticism*, Vol. XXXI, No. 3 (July, 1957), pp. 364–397; J. A. J. Peters, C.Ss.R., "Matter

to form, since it is nothing else but its absence. And so, since what is opposed to good and which takes it away is evil, it is clear that privation pertains to evil. And it follows that it is not the same as matter,[a,177] which is the cause of a thing in the way that a mother is a cause.

136(8). Then, when he says: (5), he shows the same thing through an argument leading to the impossible in the following way. Since form is something good and desirable, matter, which is distinct from privation and from form, is by its nature ordered to seek after and desire form.[b] But those philosophers who do not distinguish matter from privation are guilty of this absurdity, viz., that a contrary desires its own corruption.[c,179] And that this absurdity takes place, he shows in the following way.

If matter desires form, it does not desire it insofar as it now underlies this same form,[d] because it does not now lack being through that form (and every appetite for a thing is due to a need of something, since it does not have that thing). Likewise, matter does not desire the form to the extent that matter is under a contrary form or privation, since one of the contraries is corruptive of the other, and thus something would seek its own corruption. It is clear, therefore, that

---

and Form in Metaphysics," *ibid.*, No. 4 (Oct., 1957), pp. 479–483.

[a]Cf. St. Thomas, *In I Peri Herm.*, less. 11, n. 151, par. 2.

[b]Cf. St. Thomas, *Cont. Gent.*, Bk. III, chap. 22; cf. also Augustin Mansion, *Introduction à la physique aristotélicienne* (Paris: J. Vrin, 1946), pp. 241–281; J. A. J. Peters, C.Ss.R., *op. cit.*, pp. 447–483.

[c]Cf. St. Thomas, *De Ver.*, qu. 22, art. 1, ans. 3.

[d]"sub ipsa forma."

matter, which desires form, is as different in meaning from form as it is from privation. For, if matter by its very nature desires form, as was said,[a] and if it is stated that matter and privation are the same in meaning, it follows that privation desires form, and thus desires its own corruption.[180] And this is impossible. Hence, it is impossible that matter and privation be the same in meaning.

However, matter is also *this*, i.e., what has privation, just as if a woman were to desire the masculine and as if the bad were to desire the good, not as badness itself would seek the good which is contrary to itself, but accidentally, because that which happens to be bad desires to be good. And likewise, femininity does not seek the masculine, but that to which being feminine belongs seeks the masculine. And, too, privation does not desire to be form, but that to which privation belongs desires to be form. And this is matter.

137(9). Avicenna, however, is opposed to these words of the Philosopher for three reasons. The first of these is that neither animal appetite belongs to matter, as if self-evident, nor does natural appetite to seek a form belong to matter, since matter does not have any form or power inclining it to anything. For in this way, the heavy naturally desires the lowest place, since it is inclined to such a place by its own weight. Secondly, he objects for the reason that, if matter desires form, this is either because it lacks every form,[b,181] or because it desires to have many forms at the same time, which is impossible, or because it would be rid of the form that it has and seeks to have another, and this is also futile. It seems, therefore, that in no way can matter be said to seek form. In the third place, he objects for

---

[a]In this same par.

[b]Cf. St. Thomas, *De Pot. Dei*, qu. 4, art. 1, ans. 2.

this reason, viz., that to say that matter seeks form as the female seeks the male is characteristic of those who speak figuratively, viz., of poets, and not of philosophers.

138(10). But it is easy to resolve objections of this kind, for we should know that everything that seeks something either knows it[a] and directs itself toward it, or it tends to it from the order and direction of some knowing being,[182] as an arrow tends to a determined mark from the direction and ordering of an archer. Natural appetite,[b] therefore, is nothing but the ordering of some things to their own end according to nature. Not only is some being that is in act ordered by its active power to its own end, but also matter, to the extent that it is in potency, is ordered to its own end, for form is the end of matter. For matter to seek form is nothing but its being ordered to form as potency is ordered to act. Whatever form matter may have, it still remains in potency to some other form, and there is always the appetite to form in it. And this is not because it is tired of the form that it has,[c] nor is it because it seeks to be contraries at the same time,[d] but because it is in potency to other forms, while it has one form in act.[183]

---

[a]That which it seeks.

[b]Cf. St. Thomas, *In II De Anima*, less. 5, n. 286.

[c]In the sense that the form has, so to speak, "overstayed its time" in view of matter's capacity and appetite for another form.

[d]It does not seek to constitute contrary things together and at the same time. It does not seek to be two things at the same time.

Also, here he is not using a figure of speech but an example. For it was said above[a] that prime matter is knowable by proportion, inasmuch as it is related to substantial forms as sensible matter to accidental forms. And so, to clarify prime matter, we must use the example of sensible substances. Just as he had used the example of unshaped bronze and of the non-musical man to exemplify matter, so now, to clarify matter he, therefore, uses the example of the female seeking the male, and the bad seeking the good, for this takes place inasmuch as they contain something of the nature of matter. We must know, however, that here Aristotle is speaking against Plato, who used such metaphorical language, likening matter to a mother or to a woman, and form to a man,[b,184] and for that reason, Aristotle uses Plato's own metaphors against him.[185]

139(11). Then, when he says: (6), he shows how matter may be corrupted.

And he says that in one way it is corrupted and in another way it is not. To the extent that there is privation in it, it is corrupted when privation ceases to be in it, e.g., if we were to say that unshaped bronze were corrupted, when it ceases to be unshaped. In itself, however, to the extent that it is a certain being in potency, it is not generated and it is not corrupted. And this is clear from what follows. For, if matter comes to be, there must be some subject out of which it comes to be, as is evident from the preceding.[c] But the first subject in generation is matter, since we call matter that which is the first subject from which something comes to be in the strict sense and not incidentally, and which is in a thing after the thing has

[a]Less. 13, n. 9.

[b]Cf. Plato, *Timaeus*, op. cit., 50–51.

[c]Less. 12, n. 7; *ibid.*, middle of n. 10—end of less.

come to be. (And each of these points indicates its difference from privation, from which something comes to be incidentally and which is not in the thing after it has come to be.) It follows from this that matter would be before it came to be, which is impossible. And in the same way, everything that is corrupted is resolved into prime matter. When, therefore, it is prime matter, it is then corrupted, and thus if prime matter were corrupted, it would have been corrupted before it was corrupted, which is impossible. In this way, therefore, it is impossible that prime matter be generated or corrupted. But this does not exclude the idea that prime matter might come into existence through creation.

140(12). Then, when he says: (7), because he has already excluded the errors with regard to matter and privation, seemingly it remains for him to exclude the errors and doubts with regard to form. For, some stated separated forms, i.e. ideas, which they reduced to one first idea. And, therefore, he says that with regard to the formal principle, it is the task of first philosophy to determine whether there be one or many, and how many and what they are, and it will be reserved until that time, because form is the principle of being, and being as such is the subject of first philosophy. On the other hand, matter and privation are principles of changeable being, and this is considered by the natural philosopher. Yet, he will discuss natural and corruptible forms[a,186] in the following books of this work. And finally, he summarizes what has been said, when he states that thus it has been determined that there are principles, what they are, and how many. Now it i

---

[a]Cf. St. Thomas, *Quaestiones Disputatae De Spiritualibus Creaturis*, ed Marietti, qu. 1, art. 1, end of corpus.

necessary to make a new beginning in natural philosophy, viz., by determining the principles of the science.[a,187]

[a]*Infra*, Bk. II; cf. the summaries of the first two books of the *In Physic.* made by Pierre Conway, O.P. in "Thomistic Physics" (unpublished work, St. Mary of the Springs College, Columbus, Ohio, 1958), pp. 1–71.

APPENDIX

ıese notes are proposed to help clarify some of the terminology and meaning of . Thomas in his commentary on the *Physics* of Aristotle. Often the quotations ɔm other works of St. Thomas are given at some length, for we feel that no one n better give the meaning of St. Thomas than the man himself. When he has plained elsewhere the terms and meaning of passages in the *Physics*, it seems ise to use his own explanation. Our intention, then, is to make clear, as far as e are able, the meaning of St. Thomas.

All the translations of quotations are ours, unless otherwise specified.

esson One

In speaking of the matter and of the subject, St. Thomas means by the *matter* e *material subject*, i.e., all that can be attained by the science. It is that which e science studies. It is material being existing in nature as mobile, abstraction ade from all singularity. Actually, in the philosophy of nature, the material ıbject is any mobile being and this may be treated by another science also. The *ıbject* is the *formal subject*, i.e., the subject under that precise formal aspect which proper to this discipline. In the philosophy of nature, it is the particular degree immateriality proper to natural being which this being reveals when it is known ecisely as natural. It is mobility, the particular aspect under which mobile being studied in this science.

In short, the material subject of the philosophy of nature is any mobile being *existing in nature*, while the formal subject is the particular aspect that is revealed by mobile being *as known*, and this is the aspect of mobility.

2. This means that all knowledge is in the mind of the knower and not in things. Human knowledge is posterior to things known. According to Thomistic philosophy the forms of things known are in the knower in an intentional and immaterial way. These forms that are received by man's knowing faculty, the intellect, must be immaterial, since the intellect itself is immaterial. "Whatever is received is received in the manner of the recipient."

3. This implies, among other things, that for St. Thomas the material singular is not intelligible in act.

Before proceeding, it seems that a brief explanation of what St. Thomas means by "in potency" and "in act" is pertinent.

When we speak of being, we may consider either uncreated being (God), and this Being is pure act or pure perfection, or created beings. These latter have potency, i.e., capacity of being. For example, the essence of an angel is potency in respect to the act of existing. In other beings, i.e., in natural bodies, there is a two-fold composition of potency and act. First, the essence itself is composed of matter (potency) and substantial form (act); secondly, the total essence (matter and substantial form) is a potency in respect to the act of existing.

We have spoken above of the composition of potency and act in the lines of essence and of existence only. Now, supposing that the essence of a man is already constituted and already exists, from that very essence there follow human faculties such as the intellect, the will, and the senses. In relation to the essence from which they derive, all these faculties are as actualizations of its fundamental potentiality, yet, they are potential with regard to further individual acts.

In other words, God, because He is absolutely perfect and absolutely unchangeable is devoid of any potency or potentiality. But the being of all other things is too poor and too weak to realize simultaneously all that they are capable of being. For each one of them there is a vast gamut of possibilities. They will, however, not realize all of these. But to realize those that they do, they must *change*, but not essentially.

Potency and act, then, completely divide created beings, both in the order of substance and in the order of accidents.

Considering potency or potentiality more specifically, the capacity of being is a medium between pure non-being and perfect being. For example, as long as a thing is this and not that, it possesses the power to be that. But as long as it is this, that power which it possesses to be that remains mere power, and is not realized. Yet, this power is something real. If, for example, we consider a man asleep, he neither sees nor walks. Yet, he is not, therefore, blind, paralyzed or dumb. He is really capable of seeing, speaking, and walking. Even though he does not speak, he retains the power to speak.

If we consider being more exactly, we find that the capacity of being, though it is not nothing, is not being in the full and primary meaning of the word. And so we must find a name for this full and primary meaning of the word being as distinguished from potentiality. This is called *act* by philosophers after Aristotle. A being is *in act*, when it is somehow complete, determined and perfected. Potency, on the other hand, is the determinable and the perfectible.

Now, we are more prepared to consider why the material singular is not intelligible in act. We have already said in the preceding number of this Appendix that the known is in the knower in a purely intentional way because the intellect, as an immaterial faculty, can receive the known in an intentional, and consequently immaterial way only. Thus, what is intelligible in act is to that extent immaterial. Individual sensible matter, however, is necessary for the concrete material singular, and since matter impedes intelligibility, it is clear that the material singular is not intelligible in act. Act and form are principles of intelligibility, whereas potency and matter are by their very nature principles of unintelligibility. Matter is a source of unintelligibility because it is pure potency and cannot be a principle of activating the intellect to the act of knowing. And because of the immateriality of the intellect, things must, as we have said, be intelligible in act in order to be received, and consequently, known.

Moreover, according to Thomistic psychology, the similitude of a being is obtained from its act which is in accord with the form of the being. The similitude of a material being cannot be in the human intellect, for a material being has individual matter, and this cannot act upon us intellectually. As a composite, it is not act, and it is by its act that we know a thing. In short, since the intellect can know a thing only in its ideal or intentional similitude, which, in turn, is dependent upon the act of the thing, we cannot know the individual sensible thing directly by a purely intellectual act. In other words, we do not know the sensible thing through the intellect alone. St. Thomas insists that we know the sensible thing with the intellect, but only in union with a sense or senses.

The process by which the knowable is known or becomes intelligible is explained by St. Thomas in the *De Ver.*, qu. 2, art. 5:

> Whatever is in anything is in it in the mode of that in which it is.... Hence, our intellect, from the very fact that it immaterially receives the forms of things, does not know

> singulars.... For the one (similitude) that is in our intellect is received from the thing according as the thing acts upon our intellect, by acting first on the sense. But matter, on account of the weakness of its being, since it is being in potency only, cannot be a principle of acting. And, therefore, the thing that acts upon our mind acts only through the form. And so, the similitude of the thing that is impressed upon sense and is purified through certain degrees until it reaches the intellect is the similitude of the form only.... And because, for something to be known, its similitude must be in the knower and not in the manner in which it exists in the thing, our intellect, therefore, does not know singulars, the knowledge of which depends upon matter, because there is no similitude of matter in it....

. This division into separate sciences is made according to the ways in which iings may be separated from matter, for a thing is knowable to the extent that it an be separated from matter which is potentiality and the source of unknowability. ccording to St. Thomas, since material things are only potentially intelligible, ian needs an agent intellect which abstracts from the matter a thing may have in s objective existence and renders intelligible the form or essence of that thing. 'hus the thing becomes known through its form which has been somehow freed :om matter.

To know a thing in the strict sense is to know its definition, and definitions ary in intelligibility according to the degree of abstraction from matter found in iem. There are three different ways of knowing matter and, therefore, three ifferent aspects of matter from which a definition may abstract. These are: idividual, common and intelligible matter. It is the abstracting from these various spects of matter that constitutes the three degrees of abstraction. The following hart may prove helpful:

I. *Matter*
  A. *Sensible*
    1. Individual: the matter of Socrates, i.e., this bone, this blood.
    2. Common or universal: the matter of man, i.e., blood, bone.
    3. Intelligible. This is not sensible matter. It is the matter necessary for homogeneity in mathematical entities.

We can disregard individual matter and yet retain common matter, as, f example, when we say that man is a rational animal, sensible body is retain though universalized. In eliminating individual sensible matter from the definitic of what possess this manner of matter in reality, it is not possible to disregar common sensible matter. In the case of man, for example, this common sensib matter is of his very nature.

In the *Summa Theol.*, Ia, qu. 85, art. 1, ans. 2, St. Thomas says:

> Matter is two-fold, viz., common and signate or individual: common, indeed, as flesh and bone; individual, however, as this flesh and these bones. Therefore, the intellect abstracts the species of the natural thing from the individual sensible matter but not from the common sensible matter. Just as the species of man abstracts from this flesh and these bones, which are not in the notion of species but which are parts of the individual, as is said in *Metaphysics* VII, so also it cannot be considered without them. But the species of man cannot be abstracted by the intellect from flesh and bones.

Elsewhere in this same article this process of abstracting from individual matter explained in a rather basic manner. He says:

> The object of sense is form as it exists in corporeal matter. And because matter is the principle of individuation,

> every potency of a sense part is cognoscitive of particulars only. But to know what is in individual matter, not as it is in such matter, is to abstract from the individual matter the forms which the phantasms represent. And so, we must say that our intellect knows material things by abstracting from phantasms, and through material things considered in this way we come to knowledge of immaterial things.

The next aspect or manner of matter under consideration is common sensible atter. Some things, although they cannot exist in reality without sensible matter, ill can be defined without sensible matter. A curve is an example of such a ing. In its definition there would be no sensible matter. For a curve to be a ysical reality, however, sensible matter is necessary, whether that sensible matter the matter of a curved nose or of a curved lamp shade. In other words, a curve ay be conceived by the mind and here there is no sensible matter, but for the ysical realization and existence of the curve, sensible matter is required.

The third and last way in which matter is considered, viz., intelligible atter, is the kind that we find in mathematics. Mathematics can abstract from mmon sensible matter but not from intelligible matter, for this is the principle of mogeneity in mathematical entities. Intelligible matter is the immediate subject f mathematical quantity. It is, so to speak, its substrate. And quantity is what athematics considers. Intelligible matter is what is divisible into homogeneous arts, which parts have the same form as the whole. Briefly, and this is the nportant point, the plurality among these entities is not due to the form, for each art has the same form as the whole, but to intelligible matter. In *In VIII Metaph.*, ss. 5, nn. 1760–1761, St. Thomas says:

> And he says that matter is twofold, viz., sensible and intelligible. Now, sensible matter involves sensible qualities, hot and

> cold, rare and dense, and other qualities of this kind, but mathematics abstracts from this. But that is called intelligible matter which is understood without sensible qualities or differences, e.g., the continuum. And mathematics does not abstract from this matter.
>
> Since mathematics does not abstract from common intelligible matter, whether in sensible entities or in mathematical entities, therefore, in the definitions there must always be something as matter and something as form, as, for example, in this definition of a circle, viz., the circle is a surface figure, the surface is as matter and the figure is as form.

Intelligible matter is called intelligible for several reasons. First, even a matter, it depends upon the intellect for its very being, for its being is directl related to the intellect and not to reality. Intelligible matter is called intelligibl because it is a determinable element in relation to intelligible form. Just as in th real sensible order matter is a determinable element related to ontological and thu ultimately to the intelligible form, so also in the mathematical order intelligibl matter, the content which is determinable by mathematical form, is in the imagina tion. Secondly, it is abstracted from sensible qualities. Thus, accidents cannot b predicated of it, and for this reason it is more intelligible or more knowable Thirdly, in mathematics we have to use the imagination when we think of mathe matical entities. The imagination renders distinction. Since there must b abstraction from the sensible matter, the imagination is necessary to distinguish on circle from another circle. Not only do we have to think of a circle which does no even exist in physical reality, but we have to distinguish among circles. Throug the work of the imagination we distinguish one circle from another, but we do no know these circles as true individuals, since sensible individuals require matter an there is no distinguishing matter thus to render individuality. The imaginatio which performs this function is sometimes called the "corruptible intellect." Th nmaterial intellect does not suffice for individuation among mathematical beings, .g. circles, because, as we have said elsewhere, the object of the intellect is form. ınd the form of circle A is the same as that of circle B. Things that are not ifferent by reason of their form must be differentiated in some way and this is the vork of the imagination. Of course, St. Thomas does not mean that this is done y the imagination alone, but by the intellect working conjointly with the imagina- on. In the *De Trin.*, qu. 6, art. 1, St. Thomas says that one terminus is different ı each order of abstraction. The "terminus a quo" is always the same. This is ense knowledge. But the "terminus ad quem" varies. In metaphysics, it is in the ıtellect; in philosophy of nature, it is in the external senses, and in mathematics, : is in the imagination.

With regard to intelligible matter as a principle of individuation in mathe- ıatical beings, St. Thomas says in *In VII Metaph.*, less. 10, n. 1496 that just as ensible matter is the principle of individuation in material beings, so also is ntelligible matter the principle of individuation in mathematical beings. He says:

> But the matter that is in sensible beings, not as they are considered as sensible but as they are considered as mathematical, is intelligible. For, just as the form of man is in such matter, which is organic body, so the form of circle or of triangle is in this matter, which is a continuum or surface or body.

At the beginning of these notes on matter, mention was made of three legrees of abstraction and we have discussed only two. We have already spoken ›f those beings which need matter both for physical existence and for being known. This is what is treated in natural philosophy which studies beings which possess :ommon sensible matter. We have also spoken of those beings which St. Thomas :onsidered as needing matter for physical existence but not for being known.

These are mathematical beings and are to be studied in mathematics. There is third degree of abstraction and by this process we arrive at being that has no ne of matter either for its physical existence or for its being known. Such being studied in metaphysics. We should note, however, that metaphysics also studi material being, but from the viewpoint of being and not of matter.

The following chart may help to clarify the above divisions:

I. *Kinds of beings*
   A. Needs matter for being physically and for being known.
   B. Needs matter for being physically but not for being known.
   C. Needs matter neither for being physically nor for being known.

II. *Sciences that study the above kinds of being*
   A. Natural philosophy
   B. Mathematics
   C. Metaphysics

III. *Abstraction made in these sciences*
   A. From individual sensible matter but not from common sensible matter.
   B. From common sensible matter but not from intelligible matter.
   C. From all matter.

IV. *Degree of abstraction*
   A. First
   B. Second
   C. Third

5. There are various meanings of the words "science" and "demonstration. Here we understand the term "science" as certain knowledge gained throug

causes. By the term "demonstration," St. Thomas is referring to a "propter quid" demonstration that gives a reason why a thing is such. Demonstration is a further explanation of what is already known, since it proceeds from the more known absolutely. In this kind of demonstration there is no new discovery, i.e., no knowledge not virtually contained in the principles.

6. Sciences are specified by their objects. Just as objects differ in their being, so also will their definitions be different, since definitions express these different modes of being. We have shown before how abstraction and definition are interrelated. We have shown that we can abstract from the matter in an object only if that abstraction does not destroy the nature of that object, as, for example, we cannot abstract from the common matter in man, for then we might have a human soul, but a human soul is not a man. We cannot abstract from matter in beings that have matter in their very definition, and so, there is a connection between abstraction and definition and, therefore, between the definition and the science.

It is most interesting to note the difference in the division of the sciences resulting not only from the degree of remotion from matter but also from the approach to greater immateriality. In the *Curs. Phil.* Vol. I, Log 2, qu. 27, art. 1, p. 825, John of St. Thomas says that in the case of the act of abstraction the removal or abstraction from matter is considered from the viewpoint of the *terminus from which*, while the degree of knowability or immateriality is considered from the viewpoint of the *terminus to which*. The threefold genus of abstraction is taken from the former, while the different levels of immateriality are taken from the latter, which is the basis for the formal essence (ratio formalis) of knowability. John of St. Thomas says:

> Since the ultimate specification of the sciences is the ultimate essence of knowability, which is not further divisible, therefore, it is necessary, if the formal essence (ratio formalis) of knowability is taken from immateriality, that the ultimate and specific be taken determinately from the *terminus to which* of such abstraction, in which ultimate the abstraction comes to rest and is terminated. The specific and determinate essence (ratio) of knowability will not consist of the mere separation from matter, according as the *terminus from which* is considered, but in the ultimate determination of immateriality.

7. There is an "a priori" reason for the connection between matter and mobility. The kind of operation that a being has depends upon the kind of being that it is. A material being will, therefore, be able to act or be acted upon according to its being. Since a material thing is composed of matter and form, it will act or be acted upon as it is, i.e., as composed of matter and form. Every time a being composed of matter and form is acted upon or acts there is a passage from potency to act, and this is movement. And so, any being containing matter is mobile. Form, since it is act and not potency, does not, of itself, have the requisite for movement. (We are, of course, speaking here of the sensible order). Nor is prime matter the principle of movement, since it is potency to substantial form only.

Only material beings already substantially informed can move. Such beings are extended. The only reason that matter has mobility is that it is extended. Such matter is always in potency to a form other than the one that it has at a given moment. For there to be mobility there must be extension which involves not only matter but also form and privation. St. Thomas says in *In II Metaph.*, less. 4 n. 328:

> Whatever is moved must have matter, for whatever is moved is in potency. But being in potency is matter. And matter itself has the note of infinite, and nothingness belongs to this infinite, which is matter, because matter, as such, is understood as apart from all form.

And in n. 329 he says: "Privation is in the meaning of the infinite, for being in potency has the note of infinite only to the extent that it is in the intelligible content of privation." If it were possible to have a matter that did not involve privation, there would be no question of movement, for there would be no possibility of receiving other forms.

Although the kind of matter with which we are concerned here is not prime matter, we must not forget that prime matter is intrinsic to the consideration of natural philosophy, for, as St. Thomas says in *In VIII Metaph.*, less 1, n. 1689, prime matter "is disposed to all forms and privations in the same way that an alterable subject is disposed to contrary qualities." Whatever is changed, i.e., whatever is moved, must, therefore, have matter, whether it be changed substantially or only accidentally. (Cf. *ibid.*, 2436.) And the converse is also true, viz., that whatever is material must be mobile, for, as we have said, matter is potency to another form, and this is the foundation or basis of movement as it is studied in *In III Physic*. So also does finite form denote potency to another form, since, being finite, it does not exhaust potency.

We should also note that this potency to another form need not necessarily be in the order of generation but may also be in the order of alteration or local movement, for in the *In I De Caelo*, we are specifically told in less. 6, n. 8, that the celestial bodies have matter, but it is not the matter ordered to movement according to generation or corruption but to local movement.

8. If the subject of natural philosophy is those beings which need matter both for physical existence and for being known, it follows that "mobile being is the subject of natural philosophy." Natural philosophy, then, studies matter, i.e., sensible matter, under the aspect or formality of mobility. Matter, taken absolutely, could not be the formal subject of this science, for it could not be distinguished from the matter of mathematics which does not have movement. That the subject of natural philosophy is mobile being St. Thomas makes clear in the *In VI Metaph.*, less. 1, nn. 1155–1158. In n. 1155, he says:

> If every science is either active or factive or theoretical, therefore, it follows that natural philosophy is theoretical. Yet, it is 'theoretical,' i.e., speculative, with regard to a determinate genus of being, viz., that which can move (possibilis moveri), for mobile being is the subject of natural philosophy. And it is only with regard to 'this kind of substance,' i.e., the quiddity and essence of a thing which, for the most part, is not separable from matter in thought; and he says this (i.e., 'for the most part') on account of the intellect, which in some way is part of the consideration of natural philosophy, and yet, its substance is separable. Thus, it is clear that natural philosophy is about a determinate subject. This is mobile being. And it has a determinate mode of definition, viz., one involving matter.

In nn. 1156–1158, St. Thomas clarifies how natural philosophy differs from other speculative sciences, when he says:

> n. 1156. ...First Aristotle makes clear the mode of defining that is proper to natural philosophy. He says that in order to know how one speculative science differs from another, one must not conceal the quiddity of the thing and its 'intelligible content,' (ratio), i.e., the definition signifying it in the way that is to be assigned in each science. Investigation of the foregoing difference 'without this,' i.e., without knowledge of the manner of defining would, indeed, accomplish nothing. Since the definition is the medium of demonstration, and consequently, the

principle of knowledge, the difference in the speculative sciences must follow upon the diverse mode of defining.

n. 1157. It is to be known, however, that of those things that are defined, some are defined in the way that the snub is defined, and some in the way that the concave is defined, and these two differ, for the definition of the snub involves sensible matter, for the snub is nothing other than a curved or concave nose. But concaveness is defined without sensible matter, for no sensible body, e.g., fire or water, or any such body, is stated in the definition of the concave or the curved. The concave is said to be that whose middle is not on the same plane with its extremities.

n. 1158. All natural beings, however, are defined in a way similar to that of the snub, as is clear in the case of the parts of animals, such as nose, eye and face, as well as in the case of similar parts, such as flesh and bone. And it is also clear in the case of the whole animal. The same is true in the case of parts of plants which are leaf, root and bark. And it is also clear in the case of the whole plant. No definition of the foregoing can be given without involvement of motion. But each of them has sensible matter in its definition, and consequently, it has movement, for movement more properly belongs to all sensible matter. The warm and the cold, mingled in some way, must, indeed, be stated in the definition of flesh or of bone. And the same is true in other cases. Thus, the manner of inquiring into the quiddity of natural beings and the manner of definition in natural philosophy, viz., one involving sensible matter, is clear.

In n. 1159, St. Thomas continues to show that the natural philosopher must also consider the intellect, since it has sensible matter in its definition, which is "the first act of a physical organic body that has life potentially." Insofar as the soul is considered as such act and not as a substance separable from matter, it, too, is included in the realm of the consideration of the natural philosopher.

9. Here he does not mean that we must study metaphysics first, but he wants to set forth the common principles of this science so that these will not have to be repeated at the beginning of each division or book of natural philosophy. Here we must distinguish the procedure:

1. From the common to the less common in natural philosophy
2. From the study of the most common, i.e., being, in metaphysics to the study of the other kinds of being, as in mathematics and natural philosophy
3. In the temporal order of learning for the human being.

With regard to the first distinction we submit the following chart which follows this progression from the more to the less common in natural philosophy. This is the procedure in the order of determining, i.e., the procedure from the general to the specific:

1. The principles of mobile being as such, the science of mobile being:
   In Physicorum
2. The progression to concretion from these principles:
   De Caelo. This studies local movement. This is movement in the minimum sense for there is no intrinsic change but only extrinsic change.
   De Generatione et Corruptione
   De Anima
   De Memoria
   De Somniis
   Historia Animalium
   De Partibus Animalium
   De Sensu et Sensato
   De Plantis et Animalibus
   De Causa Motus Animalium
   De Motu Animalium
   In Meteorologicorum
3. The experimental sciences that have hypothetical premises and which proceed from these to dialectical or probable

conclusions. The progress towards the atomic species of science in the experimental or empiriological order.

4. Mathematical physics or "scientiae mediae." This is a hybrid science. Its formal element is borrowed from the second degree of abstraction.

These sciences enumerated in the first three of the above four divisions do not seem to be distinct sciences in the sense of being completely independent the one of the other. Rather, there is a progression to concretion. This cannot constitute a formal distinction between the sciences. It would seem that the philosophy of nature is a discipline which embraces the study of mobile being in general and in particular. Both the philosophy of nature and the experimental sciences study mobile being. As independent of the former, the conclusions of the latter can be but approximations to certitude. Three things can happen to a science as it moves from generality to concretion:

1. There can be strict science all the way. There is no epistemological change. This situation occurs in the case of mathematics.
2. There may be a descent into the order of probable knowledge. Here, however, the same principles of natural philosophy are used. This situation occurs in the case of the experimental sciences which proceed to a knowledge of things in their proper specificity.
3. The help of another science may be used. In this case the two sciences become a "scientia media." This situation occurs in the case of mathematical physics.

With regard to the procedure in the second of the three distinctions mentioned at the beginning of this Appendix note, i.e., the procedure from the most common, being, to the study of other kinds of being, the following chart may prove helpful.

I. *Metaphysics*:

A. *Theology*: concerns the highest grade of immateriality studied under the Light of Revelation.

B. *Metaphysics*: studies being as such.

C. *The Science of Logic*: This science of intentional being studies beings of reason as used in acquiring knowledge. Here we have a negative immateriality obtained by negative abstraction. The object studied is not real objectively, but ideal, i.e., existing only in the intentional order.

II. *Mathematics*:

A. *Arithmetic*: studies discrete quantity which is multitude. Multitude is not divisible to infinity, although by addition there is a potential infinity. Number is also a species of discrete quantity. St. Thomas tells us in *In IV Physic.*, less. 17, n. 581:

> Number may be considered in two ways, viz., as what is numbered in act or as what is numerable, as, e.g., when we say ten men or ten horses. This is called *numbered* number because it is number applied to things that are numbered. In another way number is called *that by which we number*, i.e., number taken absolutely, as two, three, four.

If number, considered absolutely, is taken to mean multitude divisible by unity, this unity is not divisible to infinity. This unity is the very principle of number. And each number of which it is a principle is a distinct species and, as such, is not itself infinitely divisible. The subject of the science of arithmetic, since it studies number, which is primarily form and not matter, is more immaterial than that of geometry which studies principally matter, i.e., intelligible matter, the matter of the continuum.

B. *Geometry*: Studies continuous quantity or magnitude that is intrinsically undetermined and which has intrinsic potentiality, while arithmetic is intrinsically definite and determined and has extrinsic potentiality. The distinction between the nature of number, which is studied in arithmetic, and magnitude, which is studied in geometry, is explained quite clearly by St. Thomas in *In III Physic.*, less. 12, nn. 390–392:

n. 390. First he gives the reasoning of what is said of the infinite with reference to the division or addition in magnitudes; secondly, he gives the reasoning of what is said of numbers by comparison with magnitudes....

n. 391. But it has been said above that there is infinite *addition* in magnitudes in such a way that through this no determined magnitude is exceeded. But there is infinite *division* in magnitudes in such a way that, by dividing, any quantity is further divided (transitur) in the direction of the less, as was shown above. He says, moreover, that this does not occur without reason because, since the infinite has the note of matter, it is contained within that which contains it, as is the case with matter. But what contains is the species and the form. Besides, it is clear from what is said in the second book that the whole has the note of form but that the parts have the note of matter. Since we proceed from the whole to the parts when we divide in the case of magnitudes, it is reasonable that in this procedure we do not arrive at any terminus which is not exceeded in infinite division. In the case of addition, on the other hand, we go from the parts to the whole which has the note of the containing and terminating form. Hence, it is reasonable that there is some determinate quantity which infinite addition does not surpass.

n. 392. Then,...he gives the notion of the infinite in the case of numbers in contrast to magnitudes. For it is said that in numbers there is some terminus in the direction of the smaller, which one does not exceed by division. But there is no terminus in the direction of the greater because in any number one can find a larger one by addition. But in magnitudes the reverse is true, as has been said.

And he gives the reason for this. First, indeed he gives the reason why in numbers there is some terminus which, by division, is not exceeded in the direction of the smaller. The reason for this is that every unity, to the extent that it is one, is indivisible, as indivisible man is one man and not many men. But any number, no matter what it is, must be reduced to the number one. This is obvious from the very notion of number, since number signifies this, viz., that there are more than one. But those that are more than one and exceed one more or less, are determined species of numbers. Thus, since the intelligible content of number includes unity and, since the meaning of unity includes indivisibility, it follows that the division of number rests in an indivisible terminus.

But what he had said, viz., that it is in the intelligible content of unity that there be more than just unity, he makes clear through the species, because two and three and any other number are derived from one. And so, it is said in *Metaphysics* V that the substance of something composed of six is this, viz., that it is six times one, but not this, viz., that it is two times three or three times two, because it would follow that there would be many definitions of one thing and many substances, because one number arises from different points of view in different ways.

(We have given the above in such length, for it is one of the clearest distinctions between the nature of number and continuous quantity and the relation between the two that St. Thomas has given us.)

III. *Natural philosophy*

A. *Philosophy of nature as such*: Studies material being and proceeds to the experimental sciences: physics and chemistry.

B. *Rational psychology*: Studies the form, the principle of life in living beings that have matter, i.e., the principle of life in the vegetative, sensitive and thinking beings, and proceeds to the experimental sciences: biology and social psychology.

The procedure in the last distinction, i.e., the procedure in the temporal order of learning, is one that concerns course curricula. Books have been and remain to be written on this subject for there are many things to consider in treating it adequately. We must consider the:

1. Nature of being to be studied
2. Nature of the human intellect
3. Chronological and mental ages of the person who will do the studying
4. Exigencies arising from the needs of the school and those of the pupils.

The difficulties in understanding may come either from the mind or from the object. In the case of metaphysics, the difficulties come from the mind since the metaphysical objects are the most knowable in themselves but least knowable to us. In the philosophy of nature, there is difficulty from the object, because cosmic beings are immersed in matter and for that reason are essentially obscure. Mathematics, however, since it abstracts from matter and motion, "shakes off" physical obscurity so that its object is more intelligible in itself than that of the philosophy of nature. Yet, since mathematics retains a necessary connection with matter, it does not offer the difficulties that metaphysics does. We are beings of sense and intellect and, therefore, find it easier to comprehend mathematics than metaphysics or philosophy of nature. The human intellect finds sensible things difficult because their study requires a great deal of experience, and it finds metaphysics difficult because this transcends the imagination and is free from all relation to matter.

It is precisely because mathematics abstracts from sensible things that are known by experience and because its object is not metaphysical being that transcends the imagination that a child may be made into a mathematician more easily than into a natural philosopher or a metaphysician. Children do not have the experience needed for the study of natural philosophy nor can they reach metaphysical things, since they do not understand them, although they may give verbal assent to them.

10. Here for Aristotle and St. Thomas, it seems that spoken language is presented as the best means of communication between thinking beings and that hearing is the sense of learning. This may seem strange to us who think of communication through symbols other than those of words that are spoken and heard. Written words, deaf mute signs, pictures, even facial expressions are symbols of the concepts that a thinking being may possess. We proceed from reality to the concept to the symbol, which is a means of communication and the only one of these three that can be communicated directly through the senses, the necessary channel of normal communication between men. Certainly, if this symbol is a word, we cannot say that it is essential for this word to be heard and that it is only accidental that it be written. Deaf mutes, indeed, communicate through symbols other than those that are heard. They communicate through sight and touch.

Now, the point at issue here is that St. Thomas and Aristotle are not speaking of communication in general. They are concerned with it in a very specific way, viz., with regard to the teaching and learning of natural philosophy. For many reasons this science was best communicated through the spoken word as delivered to listeners. Aristotle *taught* this science of the philosophy of nature; he did not first deliver it in the manner of a written treatise. This study of natural philosophy concerns movement in general and not in any ultimate specificity. It is a science that is founded upon *general* experience and does not need a laboratory observation of minute details. To comprehend what Aristotle wished to communicate, one would best be a *listener* in his class.

Since the situation is that of a teacher-student learning relationship, Aristotle refers to the sense of hearing rather than to other senses as those of sight or touch. Objects of sight, visible mobile beings, often need interpretation by a teacher. Taste and touch in their very operation are very restrictive and are much too mixed up with the object entitatively to be of great interpretive value in the order of learning. As far as touch is concerned, it may give greater sense certitude of what we touch, but this certitude is always that of the singular and is less noble than the more universal kind of knowledge that is possible through the process of learning from a teacher. Besides, Aristotle was dealing with the theory and principles of movement in general when he taught the treatise of the *Physica*. Movement in general cannot be studied through sight, taste or smell, yet we can learn with advantage by listening to what another knows about it. And this is accomplished through the sense of hearing.

It would seem that Aristotle's reason for using the method that he chose is that, in the process of imparting what he was to teach, word symbols, be they spoken, written or gestured, are preferred to other symbols. Certainly, in the realm of speculative philosophy one is dealing with word symbols as the proper and most efficient medium of communication. The goal in this kind of philosophy is knowledge that is true and in the order of human beings, written and spoken words are the media most proper to man's nature. They are signs or expressions of man's mental word, his concept. The written word, since it is a sign of the spoken word, is a symbol of a symbol and is twice removed from the "verbum mentis" or concept. The written word, since it is referred to the spoken word, may well be included in what Aristotle called "auditu."

Since Aristotle was teaching a group, however, he would better accomplish his purpose through word symbols imparted to his pupils who, in turn, could also address questions to their teacher. The history of philosophy bears witness to the achievements of the dialogue method of learning. This "question-answer" method is the most appropriate one, if we consider the nature of the subject which Aristotle and his pupils were studying. The tongue and the voice can express an indefinite, if not "infinite" number of concepts. The faculty of hearing, because it is so amenable, so passive and yet so capable, is receptive of this "infinity" and is, therefore, interpretatively receptive of the knowledge imparted to it. If we consider what we can learn from a teacher, the ear has the greatest receptive capacity.

11. Here we are told that the subject of natural philosophy is "mobile being and just that," and in n. 3 of this lesson we are told that "mobile being is the subject of natural philosophy." In the latter he is referring to all the books of the philosophy of nature and not just to the *In Physic.*, while in the former he is referring to the books of the *In Physic.*, only, for it is in these books that he treats "mobile being and just that."

12. The definitions to which he refers when he says "to understand" are those that are principles of a science, while the demonstrations to which he refers when he says "to know" are the operations of procedure in the scientific manner. It is the operation by which one knows with certitude. Accordingly, St. Thomas says that such knowledge or "science" is a result of demonstration wherein the grasp of first principles belongs to understanding and not to scientific knowledge.

13. The definition of a house that gives all the determinants or causes of the house is a complete definition and is equivalent to a demonstration, although it is not placed in demonstrative form or figure. We may demonstrate by causes. There is an order among the causes and some things, as a house may have many causes:

1. *Material cause*: the material of which the house is made. And the reason why one must have such a matter is the form.
2. *Efficient cause*: This is the reason of the formal cause. It is the agent or builder.
3. *Formal cause*: This is determined or specified by the final cause. It is the shape of the house.
4. *Final cause*: This is the end of the house, the purpose for which it is built. The end is the cause of the causality of the agent.

Actually, the purpose in view determines what material we use in order to accomplish the purpose. Thus, the order or priority of these causes goes from the final to the formal to the efficient to the material.

Now the above enumeration of determinants or causes holds true for artifacts, but in the case of natural beings, e.g., trees, there is a distinct material cause, but the other three causes become identified. The formal cause of a tree is treeness; the final cause is to have the form of a tree in its perfection, and the efficient cause is another being that has treeness.

14. All three, viz., principles, causes and elements, are causes or determinants, but they are not identical terms. We must consider causes or determinants as they are primarily principle, cause or element. The following distinctions may be made:

I. *Causes or determinants*:

A. *Element*: an intrinsic cause; that of which a thing is materially composed, e.g., the elements of man, or the letters that make up a word. An element is that into which a thing is ultimately resolved.

B. *Cause*: may be intrinsic or extrinsic. Causes do not constitute the ultimate parts. A cause is what in some way determines the being or becoming of a thing.

C. *Principle*: not a cause, although a cause is a principle. Every element is a cause and every cause is a principle, but not every principle is a cause and not every cause is an element, as a theater, in the sense that it is the point of departure of a walk, may be the principle but not the cause of the walk.

The above terminology applies to the kinds of causes used in natural philosophy.

All elements are causes in some way, but not all causes are elements. Elements are material constitutive causes. They are intrinsic in things and constitute the natural *material cause*.

Not all principles are causes. A principle implies a certain order of procession and for this reason it is possible to have a principle that is neither cause nor an element, as a point with reference to a line in mathematics. When, however, we consider a principle in natural philosophy, what is most obviously principle is the *efficient cause*.

All causes are principles. In natural philosophy the *final* and *formal* causes are the most important causes, although they are less obvious than the material and the efficient causes. The final cause is the main cause of the coming into being of things. It is the main cause of their "becoming" because it determines the agent to act, while the formal cause is that upon which the other things depend most for their "being." And so, we have the following division:

1. *Element*: material cause
2. *Principle*: efficient cause
3. *Cause*: formal and final causes.

15. Four parts of the proper meaning of the word *element* are given and explained in *In V Metaph.*, less. 4. In nn. 795–801, St. Thomas says that an element is:

1. A cause in the manner of a *cause out of which*, which is clear from the fact that element is placed in the genus of material cause, e.g., letters of the alphabet are elements of words.
2. A principle out of which something is *first* composed. Every word is first composed of letters of the alphabet.
3. It exists *in* or is intrinsic to a thing. Letters of the alphabet are within words.
4. It has a *species which is not divisible* into different species and is thus different from prime matter which has no species and from everything else material, which can be resolved into different species. Letters of the alphabet are last in resolution and first in composition.

Besides the example of letters of a word, he also gives the case of natural bodies as an example and says that mixed bodies are resolved into simple bodies.

16. This would indicate that moving and agent causes are not considered as synonymous. And the indication is true, for every cause that moves something else, but not every agent cause, is a moving cause in the sense of being a mobile cause, since there is an agent cause of being that does not itself move. God is an agent cause Who moves other beings without being moved Himself. The agent cause is a genus, while the moving or mobile cause is a species:

I. *Agent or efficient cause*:
   A. *Moving cause*: the cause of imperfect act, i.e., of motion.
   B. *Agent cause*: This can be considered in two ways:
      1. As synonymous with efficient cause
      2. As the cause of immanent action or of perfect act in neither of which is movement involved. It is in this sense that the agent cause differs from the moving cause, for it is in this sense that God is an agent cause.

17. The enumeration of the causes under the headings of principles (efficient causes), elements (material causes) and causes (formal and final causes) in this science shows how this science differs from others, for only in *this science* do we find *principles and causes and elements*.

The science of *mathematics* concerns the form only. Although we cannot posit the existence of intelligible matter, we must assume intelligible matter in the study of mathematics. In the *In I Post. Anal.*, less. 9, n. 5, St. Thomas says:

> Mathematical sciences do not demonstrate through the material cause, for mathematics abstracts from sensible matter but not from intelligible matter...and intelligible matter is considered in the manner that something divisible is considered either in numbers or in the continuum. And, therefore, in mathematics, whenever something about the whole is demonstrated through the parts, it seems that there is a demonstration through the material cause, for parts are materially related to the whole.... And, since matter is more properly said of sensible things, he did not wish to call it a material cause but a hypothetical cause.

The science of *metaphysics* investigates all the causes except the material ause, because it does not presuppose a material subject. Its subject may be aterial, but it is not essential that it be material.

8. In the ontological order, the finality of matter is the form. In the epistemological order, we must know the form in order to know the matter. If matter raws its intelligibility from the form, we cannot know the matter without first nowing what makes it intelligible. This is why in the study of natural philosophy e material elements are considered last. St. Thomas considered them in the *In Meteor*.

9. He means that we must determine which principles are more all-embracing, or the science that proceeds from fewer principles is more universal and, therefore, higher in the hierarchy of science than the one that proceeds from more. The reason for this is that a science that proceeds from fewer principles has more ertitude than one proceeding from many principles. The reason why it is more ertain is that such a science does not descend into particularities and contingencies. The knowledge of these latter is less certain than that of things considered niversally.

0. The known is, of course, known only to a knower. When St. Thomas peaks of what is more known in or by nature, he means what is more intelligible n its very nature or essence.

1. The universal meant here is the potential universal, the "universal in raedicando" which contains its subjective parts, as animal contains man and brute.

This universal is obtained through a process called *total abstraction*, which i precisely the kind of abstraction which abstracts a universal from particulars. Tota abstraction is necessary for us, for we know things confusedly before we kno them in their full determination, because, in knowing, we first know somethin considered in its potential comprehension before we know it in its ultimate ac The potentiality of a "universal in praedicando" is, therefore, proportionate to ou mode of knowing.

And so, because of the nature of total abstraction which attains the "universa in praedicando," science, for us, begins with universals, i.e., with potenti universals. These universal principles are the starting point for our science Knowledge that is confused, however, i.e., knowledge of things in general o knowledge that is indistinct, is not complete knowledge. This is why we mus proceed to specificity in natural philosophy. It does not suffice to remain in th state of generality. We must proceed to more determinate knowledge. In the orde of determining we proceed from the more general, i.e., from the confused conce obtained by total abstraction, to the specific. With regard to the order of determin ing, it would be well to read the preface to the *In Meteor.* of St. Thomas as we as the prefaces to all the treatises on natural philosophy in order to see the consta insistence upon this procedure.

Now, *the order of demonstrating*, which proceeds from principles t conclusions, or from causes to effects, has as its point of departure in natur philosophy the universal obtained by total abstraction. And so, in a science whic proceeds from universal principles we must determine the more universal one first, so that we will not have to repeat continually the general or common eac time that we approach a new specificity.

In natural philosophy the order of determining proceeds to the individual, e.g., man, from the confused universal, e.g., animal. This latter, the confused universal, is more knowable to us because it is general. We proceed to the species of movement from a consideration of movement as such. And we proceed to the study of the movement of a plant in the *De Plantis* from the study of movement that is more universal in such treatises as the *De Anima*. And so, from the examples cited here, i.e., the examples of the procedure from animal to man, from movement in general to the species of movement, and from the *De Anima* to *De Plantis*, we see in what sense we should understand the statement that "we proceed from universals to singulars."

It is interesting to note that, according to St. Thomas, there is procedure from a universal to the singular in the order of art also. In the *In II Sent.*, dist. 3, qu. 3, art. 3, ans. 1, St. Thomas, in speaking of the knowledge of the angels, distinguishes three kinds of universals:

1. "Universale in re": the universal as *singularized* in the form of the existing thing. It is the thing's nature.
2. "Universale a re": the universal *derived from* the material singular, after the manner by which we arrive at universals. It is the thing's nature but as known by the intellect. This universal is posterior to the existence of the thing.
3. "Universale ad rem": the universal *in the mind*, but this is the universal of something by nature intended to be produced outside the mind. Hence, there is the direction of the universal toward the thing. Thus, St. Thomas says in the place cited above that this universal is prior to the very existence of the thing, just as the form of a house in the mind of a builder is prior to the existence of the house in physical reality. He likens the possession of this universal to the possession of "operative science in a speculative way."

The point is that if one has a universal in the practical or operative sciences, in contrast to the speculative sciences, the former is a "universale ad rem" in contrast to a "universale a re." Specifically, in the speculative sciences we have universals received *from* nature and which we do not produce, whereas in the practical sciences, ordained to some work that we produce, the universal is ordered to production by us. Its ordering or direction is "ad rem" or toward the thing. In the *De Ver.*, qu. 3, art. 3, corpus, St. Thomas also shows that operative knowledge may be speculative.

Another point, and one that most concerns us here, is that to the extent that one moves toward operation, one moves toward the singular from the universal in the mind. Thus, one may start from "house," considered in a *universal* way, but one moves, as the plans are drawn, toward a *singular* house. It is only when one considers something operable, e.g., a house, from a purely speculative point of view, e.g., if one should consider the genus and specific difference of "house," which cannot exist as real parts of the house, that one has a "universale ad rem." It is this speculative knowledge of an operable that gives rise to St. Thomas' speaking of an "operative science in a speculative way." And to arrive at the production of this operable in its singularity, we proceed from the "universale ad rem" in the mind to the singular outside the mind.

22. "To go from what we know to what we do not know" we go from what is most knowable to us to things most knowable by nature. That we proceed from the known to the unknown is stated in a different way in the *In I Post. Anal.*, less. 1, nn. 8–12, where St. Thomas says that science is made actual in us through some pre-existing knowledge. In n. 9, referring to Aristotle, he says:

> First he presents the universal proposition that states what he proposes, viz., that the reception of knowledge in us is derived from pre-existing knowledge. And, therefore, he says *all teaching* and *all learning*, but not, however, *all knowledge*, since not all knowledge depends on prior knowledge, for then there would be an infinite regress. But the reception of all learning is derived from pre-existing knowledge. The words teaching and learning pertain to the acquisition of knowledge, for teaching is the action of the one who makes something known, while learning is the receiving of knowledge from another. Here teaching and learning are not taken solely in reference to the acquisition of sciences but in reference to the acquisition of any knowledge, no matter what it may be. This is evident since he makes this proposition clear even in the disputative and rhetorical disputations by which knowledge is not acquired. For the same reason he does not say from pre-existing *science or understanding*, but from *knowledge, generally*. He adds *intellectual*, however, to exclude knowledge in its sensitive or imaginative acceptations, since it belongs solely to reason to proceed from one another.

And in the *De Ver.*, qu. 11, art. 1, ans. 3, St. Thomas says:

> What we learn through examples we know, indeed, to a certain extent and we also do not know to a certain extent. For example, if we learn what man is, we must first know something about him, i.e., we must know the notion of animal or of substance or at least of being itself, which cannot be unknown to us. And similarly, if we are taught some conclusion, we must first know what the subject and attribute are, and also what the principles are that are known first and through which the conclusion is known. All learning, indeed, comes from pre-existing knowledge, as is said in *In Post. Anal.*

And so, we go from indeterminate to determinate knowledge in various ways.

23. Actually, volumes have been and remain still to be written on the meanings of science, certitude and probability. In the *In VI Ethic.*, less. 5, the three types of knowledge that concern necessary things are distinguished by St. Thomas. He tells us that knowledge is *wisdom* to the extent that it declares the truth about the highest causes, that it is *understanding* to the extent that it gives the truth about principles, and that it is *science* to the extent that it knows what is concluded from principles. To have certitude in the science of natural philosophy we must really know the effect by knowing it in its cause. We have a fuller knowledge of the effect through knowing the cause. Knowledge that is certain in the order of science is the result of demonstration. The conclusion must come from the principle. Of course, if we start from probable principles instead of certain principles, we will have but probable cognition. The kind of knowledge in the conclusion is determined by the kind of knowledge in the principle.

24. At the end of n. 7 of this lesson we read that "we know animal before man," indicating that we know the universal before the singular. And here at the beginning of n. 8 we read that this seems to be contrary to what is said in the *In I Post. Anal.*, less. 4, pp. 15–16, where St. Thomas says that "singulars are better known to us, but universals are better known by nature." The citation in the *In I Post. Anal.* is as follows:

> n. 15. ...To explain this, he says that what is prior and better known absolutely is what is remote from the senses, as universals are. What is prior and better known to us is proximate to sense, as are the singulars, which are opposed to universals, either by the opposition of prior and posterior or by the opposition of remote and proximate.

n. 16. It would seem that the contrary of this is held in *Physics* I, where it is said that universals are prior for us, and posterior according to nature. But here it must be said that he is speaking of the order of the singular to the universal absolutely, and this order must be considered as the order of sense knowledge to intellectual knowledge, since, for us, intellectual knowledge comes from sense knowledge. Thus, the singular is prior and better known to us than is the universal. In *Physics* I, however, the order of the singular to the universal is not taken absolutely but rather as that of the more universal to the less universal, e.g., of animal to man, and thus, for us, what is more universal must be prior and better known. In every generation, whatever is in potency is prior in time and posterior in nature. Whatever is completely in act is prior in nature and posterior in time. The knowledge of a genus is, so to speak, potential in contrast to the knowledge of a species, in which all the essential parts of a thing are known in act. Thus, in the generation of our science, we know the more common before the less common.

Likewise, in the *Physics*, it is said that the procedure from things better known to us is innate in us. Demonstration, therefore, is not attained from what is prior absolutely but from what is prior for us. We must say, however, that here he is speaking in the way that what is in the senses is better known to us than what is in the intellect. There, however, he is speaking in the way that what is better known to us is even in the intellect, for there are no demonstrations from the singulars that are in the senses, but only from the universals that are in the intellect.

Or perhaps we should say that in every demonstration we should proceed from what is better known to us, not as singulars but as universals. For nothing can be made known to us except through what is more known to us. Sometimes what is better known to us is also better known absolutely and in nature, as happens in the case of mathematics, in which by reason of the abstraction from matter, demonstrations are made from formal principles alone. In such cases, demonstrations are made from what is better known absolutely. On the other hand, sometimes what is better known to us is not better known absolutely, as

> happens in the case of natural things in which the essences and the powers of things, because they are in matter, are hidden, but are made known to us through what is externally apparent. Hence, in such cases demonstrations are made for the most part through effects, which are better known to us and not absolutely. Now, however, he is not speaking of this kind of demonstration, but of the first....

We have quoted the above reference at such length because it seemed to explain not only the seeming contradiction but also other fundamental distinctions with which we have been concerned here.

There is, then, no contradiction between n. 7 and n. 8. It is just that the word "universal" has a different connotation in the two citations. In the *In Physic.*, universals are the confused, the potential, e.g., animal, which requires further determination to be either man or brute. Singulars are the species, e.g., man or elephant, but not things in their sensible singularity as this elephant here and now. Singulars are the ultimate in the order of species, the "species specialissimae," but not a particular individual of the species. And so, here in the *In Physic.*, *singular* implies *each indivisible species* but not the individuals under that species. A singular here means the *irreducible species*. For example, in the case of Socrates, all that can be said of him "species-wise" is "man." When animal is said of him, this does not cover Socrates in his totality or in his total comprehension, for animal is only a part of him.

We should note that here both the universal and the singular are considered as objects of the intellect. We should note also that it is possible to have intellectual knowledge both of the universal and of the singular as understood in this way.

In the *In Post. Anal.*, universals are taken as proper objects of the intellect. Here animal and man are both considered as universals. These universals are more own in nature, more known in themselves than to us because of their plentitude act and their abstracting from the individuating elements that limit intelligibility knowability. Individuals, on the other hand, are considered here as objects of nse. We should note that here we can have purely intellectual knowledge of the iversal but not of the individual. We can have intellectual knowledge of the ngular but this is accomplished only through the senses. In the *In Post. Anal.* all man knowledge is considered. In the *In Physic.* only intellectual knowledge is plained and this is why the universal, i.e., the confused universal, precedes the ngular or species, for in intellectual knowledge we proceed from what is more own to us, and this is the confused or potential universal, to the study of the dividual singular, which is less known to us precisely in its individuality. ctually, we can say that natural philosophy tends toward the individual, i.e., to e singular in its irreducible species, while the *In Post. Anal.*, in its consideration, gins with the *individual*, i.e., the *sensible material singular*.

. The error of Averroes is that he expressed the "confused" as the "com- osed." St. Thomas says that the "confused" cannot be called the "composed" ecause, since genera are not composed of species, it is impossible to conclude om genera anything that gives us comprehensive knowledge of a species, e.g., of an or of elephant. The genus animal, for example, does not include the definition man but only a part of the definition of man. The interpretation of Averroes, erefore, is not correct. But this belief of Averroes occurs because, according to s philosophy, the universal subsumes the species within the limits of the concept. e believed in a separated incorruptible intelligence. This impersonal intelligence, though separated, was united to individuals in some way, for a man had a orruptible intellect, and when this man, e.g., Socrates, knew anything, this knowledge was not his own but a manifestation of knowledge that is already giv in the knowledge of a separated intellect. In this sense, the knowledge of t separated intellect is a universal, a composite, that distinctly comprises the specie And so, the most universal and the most composed are one in his theory of the tw intellects, viz., the incorruptible separated intellect and the corruptible intellect man.

26. In n. 8, St. Thomas says that genera are first known to us as containi knowledge in a confused way. If it is true that a genus is better known than t parts of the thing defined, one might wonder at the wisdom of defining a thing its essential parts. To solve this problem we must make two very importa distinctions. First, in the case of the definition of man in relation to the parts the definition, viz., animal and rational, the definition *actually* contains the defini parts, since animal and rational are intrinsic to man. Secondly, according to S Thomas, in the case of animal in relation to man and to brute, animal *potential* and not actually contains man and brute, its inferiors. Animal is better known us but in a confused manner.

In the first case, the defining elements, viz., animal and rational, are mo universal in that they are more comprehensive or more generic than the thi defined, i.e., man. There are three points to be made here.

1. The parts of the definition are more known to us because they are potential and are known confusedly. They are better known to us than the thing defined. But, and this is the important point, they are not more known *as parts* of the thing defined.
2. We must consider the knowledge of the thing defined, e.g., man. This knowledge, prior to the defining of the referent, may be the knowledge of the referent only, even if the notion

of the referent is not fully known. We may know what the noun "man" signifies, i.e., we may know the "quid nominis," without knowing the essence or nature of the thing that is signified, i.e., without knowing its "quid rei." It is in knowing the elements of a definition that we fully know the definition or essence of a thing defined. Before we know the defining elements of the definition of man, we know man only in a vague or confused way. Then, when we know animal and rational, we know man better.

3. There is the knowledge of man through its defining parts, but these parts are known only afterwards *as its defining parts*. It is in this last way that the defining parts are not known before the thing defined. And so, the seeming difficulty has been explained. We should note especially that, according to St. Thomas, we know something perfectly when we know its definition. We also know the defining element before the thing defined, although we do not know them as defining elements.

esson Two

7. Here St. Thomas says that he intends to study the "principles of mobile being such." In his summary of St. Thomas' Book I of the *In Physic.* in the *Curs. hil.*, Vol. II, p. 5, col. a, 1. 18, John of St. Thomas holds that St. Thomas says ere

> ...principles and not the causes through which natural beings come to be. He is speaking of those things from which natural being is and into which it is resolved, whether one considers natural being 'in fieri,' as it is generated, or 'in factor esse,' as it is composed of them.

n fieri" refers to natural being in the process of becoming, while "in factor esse" fers to natural being in the state of actual being.

28. The divisions given here in the first paragraph may be represented as follow

Book I: studies principles of mobile being.
II: studies principles of this science.
III: studies mobile being in general.

29. Principles are either one or many:

I. *One*
A. *Immobile*: held by Parmenides and Melissus.
B. *Mobile*: held by Diogenes (air), Thales (water), Heraclitus (fire), and others (vapor).

II. *Many*
A. *Finite*: held by those who held fire and earth to be the principles, or by those who held fire, air, and water, or by Empedocles who held that there were four principles, or by others who said that there were more than four.
B. *Infinite*: Democritus held that atoms were infinite principles. These atoms were of one genus but differed in shape, order and position. Anaxagoras held that the principles of things were infinite maximal parts. These however, are not of one genus.

We shall see later, in fact in the tenth lesson of this book, that there *is* a pluralit of principles for mobile being. If being is mobile, there must be a plurality c principles to explain the movement, for mobility implies composition and compc sition implies plurality. Actually, the tenth lesson begins the second division of thi book. The first ten lessons consider opinions and principles in a more general wa than do the remaining lessons. It is interesting to note, too, that St. Thomas studie opinions before he gives what he considers to be the truth about the principles o modern being.

30. In the seventh book of this work, St. Thomas explains more clearly what he neans by figure and the contrariety arising from it:

> n. 913. ...First, he says that from the following we should consider that whatever is altered is altered in its sensible qualities. And consequently, to be altered belongs only to whatever is substantially affected by such qualities.
>
> n. 914. ...First, he says, therefore, that in addition to sensible qualities alteration seems especially to be in the fourth species of quality. This is quality in the order of quantity, i.e., *form* and *figure*. It also seems to be in the first species of quality which includes *habitus* and *dispositions*. For it seems that there is some sort of alteration by the fact that such qualities are repeatedly removed or acquired. It does not seem, indeed, that this can happen without change. But change in quality is alteration.

In the above quotation, "repeatedly removed or acquired" means simply that from experience we see that such qualities, when present, can be removed and, when absent, can be acquired.

And St. Thomas continues:

> In the foregoing qualities of the first and fourth species, however, there is no alteration primarily and principally, but secondarily, because such qualities follow upon some alterations of first qualities, just as it is obvious that when the substratum is condensed or is rarified, change in figure follows. And, similarly, when it becomes warm or cold, there follows change in health and sickness, which belong to the first species of quality. Rare and dense, warm and cold, are, however, sensible qualities. And so, it is clear that in the first and fourth species of quality there is no alteration primarily and substantially. But the removal and acquisition of such qualities follow upon some alteration in sensible qualities.

> n. 915. ...He proves what he had assumed. And first, he proves that there is no alteration in the fourth species of quality, and secondly, that there is no alteration in the first.
>
> With regard to the first, he presents two arguments. The first of these is taken from a manner of speaking. When it is considered that form and shape differ from one another in this, viz., that shape implies the termination of quantity, that, indeed, which is enclosed by a termination or terminations, is shape. But that is called form which gives specific being to an artificial entity, for the forms of artificial things are accidents.
>
> He says, therefore, that that of which the form of a statue is made we do not call form. In other words, the matter of the statue is not predicated principally and directly of the statue. And the same is also true in the case of the shape of a pyramid or a bed. In such things, however, matter is predicated denominatively, for we speak of a triangle of bronze, wax, or wood. And the same holds true for other things. But in the case of what is altered, we predicate the attribute of the subject, for we say that bronze is humid or strong or warm. And conversely, we say that the humid or warm is bronze, equally predicating matter of the attribute, and conversely. And we say that man is white and that the white is man. Because in the case of forms and shapes matter is not called the same as the shape itself, so that the one is predicated of the other principally and directly, (but matter is only denominatively predicated of shape and form, while in what is altered, the subject and the attribute are predicated of each other), it follows, therefore, that there is no alteration in forms and shapes but only in sensible qualities.

Above, it is said that "matter is only denominatively predicated" of something. This means that the predicate belongs to a category other than that of the subject.

Returning now to the discussion of figure and its concomitant contrariety, we read:

> n. 917. ...Of all qualities shapes especially follow upon and point out the species of things. This is most obvious in the case

> of plants and animals where the difference in species can be judged by no more certain judgment than by the difference in configuration. And this is so because, just as of all accidents quantity is most closely related to substance, so also shape, which is a quality of quantity is most closely related to the form of the substance. And so, just as some people stated that dimensions were the substances of things, others stated that shapes were substantial forms.

The above quotes from the seventh book of *this work* make it clear that form and figure, understood in the sense intended, concern quantitative structure, rather than the essential form of a thing. And so, according to St. Thomas it seems that Democritus' contrariety of shape, viz., the contrariety of straight and curved is, therefore, not a contrariety arising from the essential form, but rather from the physical order. The same holds true for the other two contrarieties which he posited, viz., those of order, i.e., contrariety of prior and posterior, and of position, i.e., before and behind. The atoms, the infinite principles of Democritus, unlike the infinite principles of Anaxagoras differed in shape, order and position, but not in genus. For Democritus, whatever the ultimate differences in his principles may be, they result from a determined pre-existing difference.

31. Here St. Thomas speaks of the *beings* themselves. In the preceding paragraphs, he speaks of the principles of *being*.

32. With regard to why the ancient philosophers used only the material cause, St. Thomas in the *In I Metaph.*, less. 4, gives us the following pertinent information:

> n. 74. First, he says that many of those who first philosophized about the natures of things stated that the principles of all things were only those that are reduced to the species of material cause.

> And to say this they took four conditions of matter which seemed to belong to the notion of principle. For the principle of a thing seems to be that from which a thing is. But matter is of this nature, for we say that a material thing is made of matter, as a knife is made of iron.
>
> Likewise, that from which something comes to be, since this is also the principle of generation of the thing, seems to be a cause of the thing, because it is through generation that a thing comes into existence. But a thing comes to be first from matter, because matter exists before the thing comes to be. And also, something does not come to be accidentally from it, for a thing is said to come to be accidentally from a contrary or from privation, as we say that white comes to be from black.
>
> Thirdly, that into which all things are finally resolved through corruption seems to be the principle of things, for, just as principles are first in generation, so too, are they last in resolution. And this obviously happens in the case of matter, too.
>
> Fourthly, since principles must remain, that seems to be a principle which remains in generation and in corruption. But matter, which they say is the substance of a thing, remains in every change. But attributes are changed as form is changed and all things which are over and above the substance of matter. And from all these conditions they concluded that matter is the element and principle of all that exists.

In the above equation, we see that there are four conditions for principles, and these four seemed to be answered in the acceptance of matter as principle. For the ancient philosophers matter was:

1. *That from which a thing is*, e.g., a knife from iron.
2. *That from which a thing comes to be or is made.* This, for them, was a cause, and matter answered the question, since a thing was primarily made of matter.
3. *That into which all things are finally resolved.*

4. *What remains in generation and corruption.* Even though the forms or attributes may change, matter was the principle, since it remained.

From the fact that matter fulfilled all these necessary conditions for a principle, they concluded that it was the principle of all natural beings. Actually, the fulfillment of the first condition made matter a *principle*; the use of matter in the second condition was more as a *cause*, and in the third and fourth, it was more as an *element*.

33. For the ancient philosophers form is revealed by shape, which is a sign of species. Artificial form is the shape and the shape is the closest approach to species in the artificial order, for it is shape that terminates quantity. Because artificial things are differentiated by accidental form, we should not, however, think that artificial forms constitute the difference in natural beings. The substance of the bed and the substance of the material of the bed are the same. The shape of a bed or of a chair does not cause a change in the substantial or natural form, which is wood.

This error, which declared that natural forms were accidents, was an error of human nature, for artificial things are more known to us with regard to their proper structure, and it is in the consideration of artificial things that we learn better what we call form.

34. We might note here that since, for the ancients, matter was the only substance, first philosophy was, for them, natural philosophy, since the only substance that they knew was corporeal mobile being.

35. Although it is apparent to us and was also apparent to the natural philosophers that natural beings are moved, nevertheless, the ancient philosophers went too far in the conclusions which they drew from this. They said that since only sensible beings exist, and since there is much of the "infinite" and the indeterminate in them, because they have matter, which as such is not determined to one but is disposed to many forms, they concluded that the very being of sensible beings is not determined to one but is disposed to many different beings. And they proceeded from indeterminate sensible beings to indeterminate knowledge of these beings. In the *In IV Metaph.*, less. 12, St. Thomas says:

> n. 682. The natural philosophers do not speak the truth, when they say that there is nothing determinate in sensible beings, for, although matter as such is indeterminately disposed to many forms, yet, through form, it is determined to one mode of being. And so, since beings are known through their form more than through matter, we must not say that there can be no determinate knowledge of beings. And yet, because their opinion has some plausibility, it is more appropriate to say what they said rather than what Epicharmus said to Xenophon. Epicharmus said that all things were immobile and necessary, and were known with certitude.

> n. 683. ...The philosophers saw that all this nature, viz., sensible, was in motion. They also saw that nothing true was said of being that is always changing, i.e., of what is moved, to the extent that it is moving. For what changes from whiteness to blackness is neither white nor black insofar as it is changing. And therefore, if the nature of sensible beings is always being changed...and in all respects so that nothing in it is fixed, there is no determinate truth to be said of it. And so, it follows that the truth of opinion or proposition does not follow upon a determinate mode of being in things, but rather upon what appears to the knower so that to be true for anything is to 'appear' to someone.

It is obvious from the above citation that it is not sufficient to suppose that all movement is sensibly apparent. There is more than the indeterminacy of change. There is a "determination to one," to a given mode of being, through the form.

36. The reason for this is that science judges conclusions and not principles. It is the part of wisdom to judge principles.

37. The positions to which he is referring are those of the philosophers who considered natural beings, but who did not bring out the difficulties to which their conclusions might lead. He is referring specifically to the statements of Parmenides and of Melissus. There are several references in the *In Metaph.* which we shall cite and which make it clear that it is the task of first philosophy to argue against these positions:

> Book I, less. 4, n. 78: And it should not seem illogical if he treats here the opinions of those who dealt with natural philosophy only, because, according to the ancient philosophers, who knew no substance except corporeal and mobile substance, first philosophy had to be natural philosophy, as will be said in the fourth book.
> Book IV, less. 5, n. 593: First, he states that although none of the particular sciences should be concerned with the foregoing principles, yet, some natural philosophers were concerned with these, and not unreasonably so.

When he speaks here of the "particular sciences," he means those sciences other than the universal science of being, viz., metaphysics. And the "foregoing principles" refer to the first principles. Also, the reason why "some natural philosophers were concerned" with these first principles was that they thought matter was everything and, therefore, thought that natural philosophy was th ultimate science in their hierarchy of certitude and universality. Natural philosoph was the culmination or fruition of all that reason had attained and could attain This science was for them the embodiment of speculative knowledge. St. Thoma tells us this in the last of the above two citations, where he continues:

> The ancient philosophers, indeed, did not think that there was any substance other than corporeal mobile substance, which the natural philosopher considers. And it was believed, therefore, that they alone determine about all nature and consequently, about being, and also about the first principles which must be considered along with being. This, however, is false, because there is still a science higher than natural philosophy, for nature itself, i.e., the natural being that has in it the principle of movement, is in itself some one genus of universal being. For not every being is of this kind, since it has been proved in the eighth book of the *Physics.* that there is some immobile being. But this immobile being is superior and more noble than mobile being, which the natural philosopher considers. And since the consideration of being in general belongs to the same science to which the consideration of first being belongs, therefore, the consideration of being in general belongs to a science other than natural philosophy. And it will also belong to this science to consider the common principles of this kind. The *Physics*, indeed, is a part of philosophy, but it is not first philosophy which considers being in general and what belongs to being as such.

And in *In VI Metaph.*, less. 1, n. 1170, he says:

> If there is no other substance except those that exist naturally, which physics considers, then physics will be first science. But if there is any immobile substance that will be prior to natural substance, consequently, the philosophy that considers substance of this kind will be first philosophy. And since, if it is first, it will, therefore, be universal, and it will be

> its function to speculate about being as such and about its nature (de eo quod quid est), and about what belongs to being as such, for the science of first being and of being as such is the same science, as is considered in the beginning of the fourth book.

St. Thomas repeats the same theme in *In XI Metaph.*, less. 7, n. 2267.

> If natural substances, which are sensible and mobile substances, are first among beings, natural philosophy must be first among the sciences, because there is an order in the sciences and it corresponds to the order of subjects.... If, however, besides natural substances there is another nature and substance that is separable and immobile, there must be another science of this. And this science is prior to natural philosophy. And from the fact that it is first, it must be universal, for the science that concerns first beings is the same science that is universal, since first beings are principles of others.

In the above quotations we see repeated the following two points:

1. The metaphysician must also consider sensible beings, since he considers all substance.
2. If all substance were sensible, the philosophy of nature would be metaphysics or the most universal science.

Lesson Three

38. Being is an analogous term. Here Parmenides erroneously takes it univocally instead of analogously.

39. To understand the meaning of "first and second substance" some references may be helpful.

In the *Summa Theol.*, Ia, qu. 29, art. 1, ans. 1, St. Thomas says, "...grante that this or that singular cannot be defined, still, what belongs to the commo notion of singularity can be defined, and in this way the Philosopher defines fir substance." And in ans. 2, he says, "...substance is taken in a general sense, a divided into first and second substance; and by the fact that the individual is added it is taken as standing for first substance."

The explanation of the meaning of first and second substance seems cleare in the *De Pot. Dei*, qu. 9, art. 2, ans. 6, where St. Thomas says:

> Since substance is divided into first and second substance, it is not a division of a genus into its species, since there is nothing in the second substance that is not in the first. It is rather a division of genus according to the different modes of being, for second substance signifies the absolute nature of genus as such, while first substance signifies it as individually subsistent. Hence, it is a division of an analogue rather than of a genus.

And since he is speaking here of the nature of the person, he adds, "In this way therefore, the person is, indeed, contained in the genus of substance, although no as a species but as determining a special mode of existing."

John of St. Thomas comments upon this quote from the *De Pot. Dei* in hi *Curs. Phil.* Vol. I, qu. 15, art. 2 of the second part of his treatise on logic, where he is commenting upon the *In Peri Herm.* (In the Reiser edition this is pp. 530 ff.) After presenting the quote from St. Thomas, John of St. Thomas says:

> And from this his followers know that St. Thomas did not feel that it was an analogous division, but that it was more a division of an analogue than of genus, although it is neither, but rather the division of a subject or of a nature into modes of being. This is likened to the division of an analogue that is made through modes and not through differences, in the way a genus is. Finally in the *Ia*, qu. 29, art. 1, ans. 2, St. Thomas says that in the definition of the person, when it is

said that it is an individual substance, there substance does not determinately signify first substance but 'what is divided generally into first and second, and by the fact that the individual is added, it is taken as standing for first.' According to St. Thomas, therefore, what is stated in the name of substance in the definition of the person is what is divided into first and second substance. But when the person is called an individual substance, it is not the same as the individual property of subsisting, but it is the substance itself. This, therefore, is the way by which this division is made. And so, because the first substance itself is not only a determined property of subsisting but also a determined thing that has the property of subsisting in the first manner of 'per se' predication, it is more suitably said that this division is not one of property into properties but of the thing that has properties into the modes of such properties.

From this it is established that first and second substance do not mean different natures, since they are related as singular and universal in the same predicament, and in this way we can say that to subsist in the first way and to subsist in the second way is the same, because it is the same nature. Those modes or properties cannot, however, be distinct realities, since no real property that does not belong to singular substance may belong to universal second substance, but that very notion of real subsisting, which is found in the singular, also belongs to the universal nature. But the activity of subsisting or of sustaining accidents belongs really to first substance and nominally to second....

And so, we have established the kind of division that this is. If we understand that that which is divided is the nature of substance as affected by those two modes, there is, then, a division of the thing into its modes or of the subject into its accidents, as we have seen from St. Thomas' question in the *On the Power of God*. This does not contradict what St. Thomas said, viz., that the division was that of an analogue into its analogates rather than a division of a genus. Saying that the division is analogous is not, however, speaking absolutely, but he means that in contrast to the division of genus, which is a division into different species and natures, that division is closer to the division of an analogue than of a genus, which is made through different modes, one of which, viz., to subsist in the first way (primo), is really found in first substances, but to subsist in the second way (secundo) is found only nominally and notionally (per rationem) in first substances, as if nature were divided into universal and singular. But if we understand that the notion of this property or mode of subsisting is divided, then the division is analogous to the extent that one mode, viz., to subsist in the first way (primo) is really found, but the other, i.e., to subsist in the second way (secundo) is found only nominally and notionally.

(Reiser edition erroneously refers to qu. 19 instead of 29.)

Briefly, then, we may summarize the above in the following distinctions:

1. *First substance*: individual: this man: individual substance: subsists in the first way (primo).

2. *Second substance*: universal: man understood universally: universal substance: subsists in the second way (secundo).

Both first and second substance are substance, however, and are, therefore, of the same nature. For this reason the division is likened to a division of an analogue. Cf. Appendix note 69.

40. The sense in which we use the meaning of infinite here is found only in the quantified, for we are taking infinite as that of which there is always something beyond, as we find in the study of the infinite in the third book of *this work*. This definition is specifically stated in n. 383 of that book. The Aristotelian and Thomistic procedure from quantity to the infinity that is proper to quantity indicates

a multiplicity which was obviously against the belief of Melissus for whom being was not multiple but one.

In the sensible order of quantity, we cannot say that anything is either finite or infinite unless that thing be quantified, since it is quantity that makes possible finitude or infinity in this order.

41. The act of the continuous is one and makes abstraction from the divisibility of the continuous. To be one as the continuous is one is not to be absolutely one but to be potentially multiple, for, although the continuous is not divided, even though it has parts that are in potency to be actual parts, it can be divided. Since the continuous is extended, and since to be extended is by definition to have parts outside of parts, the continuous, according to St. Thomas, has parts that are actually and not merely potentially parts, but these parts are not actually separated. The nature of the continuum is what gives it the potency to actual parts. If we abstract from the possibility of divisibility to infinity, we no longer have a continuum, for its nature then, is denatured.

42. He is speaking here of qualities that follow upon quantity. There are, however, some qualities that are not in quantity, as that of having knowledge. This is a quality of the intellect, and the intellect is not quantified.

43. Infinite, of course, is understood here in the physical sense.

44. Contraries are extremes in the same genus. White and black are extremes in the genus of color. Contradictories, as the name implies, are not opposites in the same genus. One contradictory is the negation of the other; e.g., sitting and not

sitting, book and non-book are contradictories. Book is not in the genus of non-book, nor is sitting in the genus of non-sitting. (Cf. *In X Metaph.*, less. 6, nn. 2040 ff.)

Now, if all things were one in meaning, it is said that "contraries would be one in meaning," which is absurd. The reason why this is so is clarified by St. Thomas in the *In IV Metaph.*, less. 15, n. 719, where he says:

> Because it is impossible that contradiction be verified of the same thing at the same time, it is clear that neither can contraries be in the same thing, for it is obvious just as much in contraries as in other opposites that one of them is a privation, even though each of the contraries is a nature. And this is not so in the case of affirmation and negation or in privation and possession, for one of them is imperfect with respect to the other, as black with respect to white and bitter with respect to sweet. And so, there is a kind of privation implied in it. But privation is 'a kind of negation of substance,' i.e., in some determinate subject. And it is also within a determinate genus, since the negation is within the genus; for not every non-seeing thing is blind, but only what is in the genus of the seeing. And so, it is clear that a contrary includes privation, and that privation is a kind of negation. If, therefore, it is impossible to affirm and deny simultaneously, it is impossible that contraries be absolutely in the same thing at the same time, but either 'each is in it in the same manner,' i.e., in some respect, as when one of the two is in a thing in potency or in respect to a part, or as one is relatively in a thing and the other is absolutely in it, as when one is in act and the other is in potency. Or one is in a thing in respect to more parts and more principal parts, and the other is in it only with respect to some part, as, for example, an Ethiopian is black absolutely and white with reference to his teeth.

45. St. Thomas says in the *In IV Metaph.*, less. 6, n. 606:

> Some, as they say of Heraclitus, said that the same thing happens to exist and not exist at the same time, and that men happen to believe this. And many natural philosophers use this position, as will be clear below. But we now assume the truth of the foregoing principle, viz., that it is impossible for the same thing to be and not to be. But basing ourselves on its truth we show that it is most certain. For from the fact that it is impossible to be and not be, it follows that it is impossible for contraries to be in the same thing at the same time.... And from the fact that contraries cannot co-exist, it follows that man cannot have contrary opinions, and consequently, he cannot think that contradictories are true....

And in n. 737 of less. 17 of the same book, he says:

> Some said that nothing is true but that all things are false, and that nothing stops us from saying that all things are false in the way that this is false: the diameter is commensurate with the side of a square. But others said that all things are true. And such statements are a result of the opinion of Heraclitus, as has been said, for he spoke of being and not being at the same time, and consequently, all things are true.

ł6. Just as there is an intelligible content of being and of good, so also is there ɔne of non-being and of evil; for everything definable has an intelligible content. Non-being has a proper intelligibility; otherwise, we could not compare it to being. We must also suppose that the meaning of good and that of evil are not the same ntelligible content. While the opposition of being and non-being is one of ɔrivation, the opposition of good and evil is one of contrariety. Certainly, being and non-being are not opposites in the same genus, for being has no opposite. It s an all-embracing genus.

Evil is not mere privation. It is the opposite of good and is in the same genus as good, but since the good and the evil are always a good something and a bad something, good and evil as such are not really contraries, but are implied such.

In *In I Sent.*, dist. 19, qu. 5, art. 1, ans. 8, St. Thomas, in discuss whether or not truth is the essence of a thing, says:

> Being is the first intention of the intellect; hence, nothing can be opposed to being by way of contrariety or of privation, but only by way of negation. The reason for this is that just as being itself is not founded upon anything, neither is its opposite; for opposites regard the same thing.
>
> But one, true and good, according to their proper intentions, are founded upon the intention of being, and, therefore, cannot have the opposition of contrariety or of privation founded upon being, since they themselves are founded upon being.
>
> Thus, it is clear that the true and the false and the evil and the good are not related in the same way as being and non-being, unless non-being is taken as a particular for the removal of something of which a being is capable. And so, just as any privation of particular being is founded in the good, so also is the false founded in something true as in some being. Hence, just as that in which there is falsity or evil is a being, but not complete being, so also that which is evil or false is *a* good or incompletely true.

In short, being has no contrary, but only a contradictory; the good and evil are opposed as contraries and not as contradictories, but their opposition contrariety is actually only an implied one, insofar as there is a good something a a bad something, a concrete good and a concrete bad. But insofar as the evil, i. a privation of the good, is founded in the good as in a being, there is no r contrariety, for contraries are not founded upon one another.

7. The reference here is to the sixth lesson where, after discussing the first rinciple of demonstration, which is that it is impossible for the same thing to be nd not be at the same time and in the same respect (cf. n. 600), he says that some vanted to demonstrate this self-evident indemonstrable first principle. He tells us nat metaphysics could not demonstrate its principle absolutely, since there would ave to be a procedure to infinity. And this is impossible. Its first principle, herefore, must be self-evident. We cannot demonstrate the first principle by other rinciples, if an adversary denies the first principle. He says, however, that this rinciple may be demonstrated argumentatively or by using an elenchus, which is ot a demonstration absolutely, but only a refutation in syllogistic form. Here St. 'homas says that an elenchus is a "syllogism for refuting."

The demonstration strictly speaking and the elenchus are distinguished more learly in n. 609, where St. Thomas says:

> Demonstrating the foregoing principle absolutely is different from demonstrating it argumentatively or elenctically; for if someone wanted to demonstrate the foregoing principle absolutely, he would seem to beg the question, because there is nothing that he could take for its demonstration except when depended upon the truth of this principle.... But when a demonstration is not of this kind, i.e., absolute, then there is argumentation or an elenchus and not demonstration.

And in n. 610, he continues:

> Another text...is better: 'But when there is a reason for this other, there will be argumentation and not demonstration," i.e., when such a procedure from what is less known to this more known principle is made the reason why the other person denies this, then, it will be possible to have argumentation or elenchus, i.e., a syllogism contradicting him, but not demonstration, since what is less known absolutely is conceded by the

> opponent, and thus it will be possible to proceed to showing the foregoing principle as far as he is concerned, although not absolutely.

48. We should note here that all the distinctions that are used are less know absolutely because the position of the adversary supposes confusion on secondar points that are anterior to the understanding of the first point.

Here the first point is that "being is one," and it is a point to be refuted. Bu instead of attacking an adversary on this point considered in itself, Aristotle, an St. Thomas after him, break down the argumentation to remove confusion o secondary points that are anterior to the understanding of the first point. Fo example, they show from the viewpoint of being that being has many meaning They continue to be make it clear that in no way could Melissus posit only on kind of being. And so, for Melissus, being could in no way be one only.

Then they show that being cannot be one from the viewpoint of unity. An they say that something can be one in three ways, i.e,. as the continuum is one c as the indivisible is one or as there is a unity in meaning. And they show tha being cannot be one in any of these three ways. Thus, Aristotle and St. Thoma have made clear that which is less known absolutely to establish their point, viz. that being is not one but many.

Lesson Four

49. An incomplete meaning would indicate that there is neither truth nor falsit and consequently, there is not even probability.

). "To whiten" is not a thing or "res." It is not something determined. It is ›ither white nor non-white so long as it is the very process of becoming white. ı movement, while there is a passage from potency to act, there is neither the first ›r the second terminus but some intermediate state. This notion "to whiten" ›comes very important in the third book of the *In Physic.* where St. Thomas udies the nature of movement. To become white is an activity of a thing in ›tency to be moved. Neither white nor moved can be called a "res," although ey may be in a "res." This may be somewhat clarified by the example of the act f building. In Bk. III, less. 3, n. 2, St. Thomas says:

> Everything that is in potency may at some time be in act. But the buildable is in potency. There may, therefore, be some act of the buildable insofar as it is buildable. But this is either a house or the process of building. But the house is not the act of the buildable as buildable, because the buildable as such is reduced to act when it is being built. It remains, therefore, that the process of building is a certain motion. Motion, therefore, is the act of a being in potency insofar as it is of this nature....

1. What is said here about the one and the many shows that there is confusion ecause of the failure to understand the two problems of the one and the many. he two problems are the *objective* one taken from the viewpoint of the object in self, the natural problem, and the *noetic* problem that concerns the one and the ıultiple from the viewpoint of the multiplicity of the means of knowing. armenides and his school have confused these two problems. We often admit 'hat he says with regard to his noetic one and multiple, but with regard to his atural one and multiple what he says is generally unacceptable.

In the natural order the same thing may be the subject both of whiteness and f whiteness in the process of becoming corrupted. There is a unity and a multiplicity, for there is a unity in subject and a multiplicity in meaning. In the noetic order, because there is a unity that we cannot attain in some objects considered in their very distinction and diversity, we try to surmount and surpass our means of knowing. This, however, is impossible for us. We need a different concept for each kind of thing known. We do not know many things in one intelligible species, i.e., we do not know many things distinctly in one concept. In addition to the concept of a thing, man needs judgments to really know a thing, whereas the angels can grasp the whole thing intuitively. Since the angels, according to St. Thomas, are superior to us in being, they can know many things in one act of knowledge. They achieve a greater noetic unity from the objective multiplicity of impressions that go to make up one intellectual view. In the case of man, he can know the essence of a thing conceptually, but all the impressions that he has of a thing he must know through judgments. No existence whatever is expressed in conceptual knowledge but only through judgment. The existence of a thing known as well as the existence of certain qualifications and impressions of the thing known are known through judgment. This is not the case with the angels.

In the *Cont. Gent.*, Bk. II, chap. 96, St. Thomas clarifies the kind of intuitive, all embracing knowledge that an angel may have of a thing or things. He says:

> The object of a higher power must be higher. But the intellective power of a separated substance is higher than the intellective power of the human soul, since the intellect of the human soul is lowest in the order of intellects.... Moreover, the object of the human soul is the phantasm...which is higher in the order of objects than the sensible thing existing outside the soul, as is evident from the order of cognoscitive powers. The object of a separated substance, therefore, cannot be a thing existing outside the soul, as that from which it would immediately

> receive knowledge; nor can it be a phantasm. It remains, therefore, that the object of the intellect of a separated substance is something higher than the phantasm. But in the order of knowable objects, nothing is higher than a phantasm except what is intelligible in act. Separated substances, therefore, do not derive intellectual knowledge from sensibles, but they know those things which are intelligible even through themselves.

And in chap. 98, he says:

> Consequently, in the case of higher substances, knowledge gained through more universal forms is not more imperfect, as it is with us. For through the similitude of *animal*, by which we know a thing in genus only, we have more imperfect knowledge than through the similitude of man by which we know the complete species, for to know a thing in its genus only is to know it imperfectly and as in potency, but to know a thing in its species is to know it perfectly and in act.
>
> But our intellect, which has the lowest level in intellectual substances, requires similitudes so particularized that for each proper knowable it must have in itself a proper corresponding similitude. For this reason through the similitude of *animal* it does not know *rational*, and consequently, it does not know man, except in some respect.
>
> On the other hand, the intelligible similitude, which is in a separated substance, is one of more universal power and is sufficient for representing more things. Hence, it produces not more imperfect but more perfect knowledge, for it is universal in its power, in the manner of an agent form in a universal cause, which, to the extent that it is more universal, is extended to more things and produces more efficaciously. Consequently, through one similitude the separated substance knows both animal and its differences: or it even knows them in a more universal and more contracted way, according to the order of these substances.

And so, something known by multiple means is always known imperfectly, or it is not known in its intense rigorous unity. As long as we have to employ

manifold means of knowing, our knowledge, while it may be true, is not perfec
It is not the multiplicity of objects in question but the multiplicity of the means c
knowing that makes for imperfection in knowledge. Knowledge, then, is nc
necessarily imperfect when the things known are multiple, but something known b
multiple means is known in an imperfect way.

In trying to illustrate unity in both the noetic and real orders, if we conside
"man whitens" and "man is white" from the viewpoint of these two orders, w
find that from the viewpoint of the objective or natural or real order, the probler
of the one and the many takes on a kind of unity in the expression "man whitens.
It is even possible here that "whitens" refers not so much to the very activity c
becoming white as to an activity of man. The expression "man whitens" i
intended as a surrogate of the expression "man is white" and of the expressio
"man is becoming white." "Man whitens" expresses that being white is an activit
of man in the sense that when man whitens, he does what a white man does, ju
as the expression "light shines" makes known the activity of the light, which doe
what a light does.

There is also the possibility of explaining this objective unity in th
expression "man whitens" on the basis of the nature of the process of becomin
The man is not definitely white or black. There is an indefiniteness. From th
standpoint of natural unity, this seems to be better than the expression "man
white," which is very definite and does not allow for the unity of the continuity c
process. But even the expression "man is white" has natural unity, for there is bu
one substance and this happens to be white in color.

From the viewpoint of the noetic one and many, however, the notions of ma
and of white or of becoming white cannot be reduced to one concept by the huma

tellect. The intellect's frustrating unifying tendency cannot reduce everything to identity. It would be a tendency to explain all in terms of a tautology.

esson Five

2. The argument of Melissus is not defective, because his argumentation is not ven an argument. It sins against the very order of natural philosophy, for the gument assumes what is contrary to that order. It is not defective in the logical der, since it proceeds logically from what is obviously false. But since it offers othing for argument to the natural philosopher, there is no problem. The natural ilosopher cannot argue with or attack a man's argument if there is not an gument.

According to St. Thomas, the argument of Melissus says that "what is made as a beginning." No natural philosopher would argue with this, because the atural philosophers say that things come to be from something, and thus there is "beginning to be" for some new thing and perhaps "a beginning to be a process" r some new thing. The second statement of Melissus stated in this lesson is that what is not made does not, therefore, have a beginning." But when Melissus says being is not made," i.e., being is not the result of some coming into existence, is is contrary to natural philosophy; for the natural philosopher says that for being be it must be some being. If it is some being, it must have had a beginning. It ust have some principle or principles of coming to be.

The subsequent statements of the argument of Melissus follow logically from is error, viz., that being is not made nor does it come to be. In this way, his rgument is not defective logically nor does it raise doubts because of its manner f logical statement. It is, however, "irksome," as far as the natural philosophers are concerned, for they hold that being must come to be and so must come to b from something. Thus, being, and here they mean material being, has a beginning Being, for the natural philosopher is, therefore, mobile, but Melissus goes on t posit the contrary of this, viz., that being is one, immobile and infinite. It is no therefore, the task of the natural philosopher to argue against statements that ar obviously false and contrary to his major tenets.

53. Here in this lesson, St. Thomas says that Melissus took the meaning c beginning in one way only, viz., as the beginning of extension. But St. Thoma says that beginning has two meanings, viz., the beginning "of time and c generation" and the beginning "of the thing or of extension." The beginning c whatever has a beginning is not necessarily the beginning of a thing's extension The term beginning has a variety of connotations. It may mean the beginning t be of a process, or it may mean just beginning to be. Man, for example, begin to be, but he does not begin in the sense of proceeding from the less perfect to th more perfect. Man begins to be from a temporal beginning. His beginning is nc the beginning of extension.

In the *In V Metaph.*, less. 1, pp. 750–760, St. Thomas further clarifies th meaning of the term beginning:

> n. 751. It should be known that principle and cause, although they are the same in subject, differ, however, in meaning; for this word principle implies a kind of order, but the word cause implies a kind of contribution (influxum) toward the being of what is caused. The order of priority and posteriority, however, is found in different ways, but according to what is first known to us there is the order found in local motion, since that motion is more obvious sensibly.

> There are, moreover, orders consequent upon three things, viz., the orders of *extension*, of *motion*, and of *time*; for, according to priority and posteriority in magnitude, there is priority and posteriority in motion; and according to priority and posteriority in motion there is priority and posteriority in time, as is considered in the fourth book if the *Physics*. Because a principle is said of that which is in some sort of order and because the order which is considered according to priority and posteriority in quantity is prior in our knowledge, and, furthermore, because they are denominated by us according as they are known to us, therefore, if we take the order of inquiry which is proper to it, this word principle signifies that which is first in *extension* over which motion passes. And, therefore, he says that that 'from which something first moves a thing,' i.e., some part of magnitude from which local motion begins, is called a principle. Or, according to another text: 'That from which something of the thing will be moved first,' i.e., from that part of the thing from which something first begins to be moved, is called a principle; just as in the case of length and any path, the beginning is from that part from which the motion begins. But from the opposite or contrary part, it is 'different or other,' i.e., the end or the terminus. It is to be known that the beginning of motion and the beginning of time in the foregoing meaning are proper to this kind of principle.

Now, St. Thomas continues his investigation of the kinds or meanings of principle or beginning and makes it clear that, contrary to what Melissus held, there are more meanings of beginning than the beginning of the thing.

> n. 752. But because *motion* does not always begin from a beginning of extension but rather from that part from which each thing is on the point of moving, therefore, he posits the second kind when he says that the principle of motion is said in another way as 'that from which anything from which each thing for the most part will best come into being,' i.e., from which each thing best begins to be moved. And he makes this clear by a simile, viz., in the case of disciplines in which someone does not

> always begin to learn from what is the beginning absolutely and as such but from that from which someone 'can learn' more easily and more readily, i.e., from those things that are more known to us and which sometimes are posterior in nature.

Thus far, St. Thomas has clarified two kinds of beginning, that of *extension* and that of *motion*, and he gives as an example of the latter the order or procedure in learning from things that may be posterior in nature but more known to us.

> n. 753. But this kind of beginning differs from the first; for in the first kind, the principle of motion is designated from the beginning of extension. But here the beginning of extension is designated from the principle of motion. And, therefore, even in those motions which trace circular paths, which do not have a beginning, there is understood some beginning from which the mobile is best or readily moved according to its own nature; just as in the case of the first mobile being, the beginning is always from the East. Even in our motions, man does not always begin to move from the beginning of the path but sometimes from the middle or from some other terminus from which it is opportune for him to move first.

The above quote makes it clear that in the second connotation of beginning the notion of priority and posteriority in extension depends upon priority and posteriority in motion.

> n. 754. But the order which is considered in local motion makes known to us the order in other motions. And, therefore, these other motions follow upon the meanings of beginning which are taken according to beginning in the generation or the becoming of things. This principle, indeed, is disposed in two ways; for it is either 'existing in it,' i.e., it is intrinsic, or, it is 'not existing in it,' i.e., it is extrinsic.

> n. 755. In the first way, therefore, that part of a thing which is generated first and from which the generation of a thing begins

> is called a principle, e.g., in the case of a ship, first the bottom or keel is made, and this is as the foundation of the ship and all the other boards of the ship are connected to it....
>
> n. 756. Principle is said in another way as that from which a thing's coming into being begins. But this principle is extrinsic to the thing; this, indeed, is classified in three ways: *first*, in the case of natural things in which that from which motion naturally begins in those things which come to be through motion is called a principle of generation, as in the case of those things which are acquired through alteration or through some other motion of this kind. So also, man is said to become great or white. Or, principle is called that from which transmutation begins, as in the case of those things that do not come to be through motion but through change alone, as is evident in the coming to be of substances, as a boy is from the father and the mother who are his principle, and as war is from wrangling which arouses the minds of men to war.

The third meaning of principle, viz., that of a *temporal beginning*, has been given with examples. We see that priority and posteriority in time is dependent upon the same order in motion and, in the natural order, motion is impossible without extension.

> n. 757. *Secondly*, it is also clear in matters of action, whether moral matters or political ones in which the one by whose will or intention other things move and change is a principle. And so, those are called rulers (principatus) in states who attain power and command, or, even tyrants in these states are called a principle; for from their will all things in the states come to be and move.
>
> n. 758. He states the *third* example in the case of artificial things, since in a similar way arts are called principles of artificial things, because the movement to the construction of artificial things begins from art. And among these arts, the architectonic arts are most called principles....

In nn. 759 and 760, St. Thomas shows that some things can be called principles because of their similarity to the order considered in extrinsic movements. In n. 759, he cites the order in coming to know things, since the intellect discourses from principles to conclusions and also, since there is a point from which we begin to know a thing. And St. Thomas speaks of the postulates of a demonstration as principles. In n. 760, he says that in the same way, i.e., by similarity, "causes are also called principles.... For the motion to the existence of a thing begins from a cause, even though cause and principle are not the same in meaning."

In n. 761, St. Thomas reduces the three kinds of principles, viz., of extension, motion and time to a common denominator. They are all *first*. And he continues to say what seems very similar to what he has said of principle in the *In Physic.* when dealing with Melissus and his misunderstanding of principle. He says that these principles are *first*

> either in the existence of a thing, as the first part of a thing is called a principle, or in the becoming of the thing, as a first mover is called a principle, or in the case of the knowledge of a thing.

And in n. 762, he explains that, although principles all agree in that they are *first*, some principles are intrinsic and some extrinsic, as he had mentioned before.

Briefly, then, we may conclude from the foregoing that Melissus was wrong in thinking that principle had but one meaning. St. Thomas has pointed out that there are three kinds of beginning:

1. *Principle of extension*. This is some part of an extension or magnitude, e.g., a piece of string.
2. *Principle of motion*, e.g., a man's point of departure for a trip, or the moving of the foot of the first horse in the race.

3. *Principle of time*, e.g., the first point of departure in learning something, or the beginning of the process of the race.

nd he reduces these to one common denominator, viz., that of "firstness," either
:

1. *The beginning of the existence* of a thing, i.e., the fact of beginning.
2. *The beginning of the becoming* of a thing, i.e., the beginning to be a process.

I. Even though the world would be taken to be infinite in size, this would not event the parts within the world from being moved locally, temporally or any her way.

esson Six

5. One might ask why, for Parmenides, being could not be infinite in the way at Melissus called matter infinite. For Parmenides there was nothing beyond. ccording to him, all things, even those that diversify, are being, and so, all was onsidered to be being in the same sense. In the *In I Metaph.*, less. 9, n. 138, we e told that Parmenides

> seemed to touch on unity according to reason, i.e., from the viewpoint of form, for he argued as follows: Whatever is other than being is non-being. And whatever is non-being is nothing. Whatever, therefore, is other than being, is nothing. But being is one. Whatever, therefore, is other than one is nothing. In this it is clear that he was considering the very principle of being (essendi) which seems to be one, because he cannot mean that anything that diversifies might be added to the notion of being, for what is added to being would have to be extrinsic to being.

> But what is of this nature is nothing. And so, it does not seem that it can diversify being. In like manner, we also see that differences added to a genus diversity it, even though the differences are outside its substance, for the differences do not share in the genus, as is said in the fourth book of the *Topics*. Otherwise, the genus would be of the substance of the difference, and there would be foolishness in the definitions if, after the genus is stated, the difference is added—if the genus were of its substance—just as it would be foolishness if the species were added. Also, the difference would in no way be different from the species. But what is outside the substance of being must be non-being, and so, this cannot diversify being.

If being were a genus, its differences would have to be other, i.e., no being. But a difference which is non-being would be a non-difference. Bei cannot be a genus, since, in order to be so, the difference would have to *a* something, and this is impossible, since it would have to be some form of bein too. If the difference did not add anything, then the genus would be the san substance or essence as the difference.

St. Thomas continues in n. 139, where he says:

> But they were deceived in this, because they used being as if it were one in meaning and nature, as is the nature of a genus. This, indeed, is impossible, for being is not a genus but is said of different things in many ways. And, it is, therefore, said in the first book of the *Physics* that this is false, viz., that being is one, for it is not one in nature, as is one genus or one species.

56. That there is only one being is the old Parmenidean problem. Plato presen this argument in detail in his dialogue *Parmenides*, where, at the request Socrates, Parmenides attempts to prove the points of his argument.

This problem is also one for those who hold the theory of the transcendentals ccording to which of whatever one of the transcendentals may be predicated, so so may the others. To say that Parmenidean being is "one" in the way that the anscendentalists mean "one" would be superfluous, for there would be nothing se by comparison with which it could be "one" and there would be nothing to ivide or separate it. All that could be said of being, which, according to armenides, is precisely a being, would be that it is, and that it is one, true and ood. There would be no degrees of being, oneness, truth or goodness with which r by which it might be compared. To the extent that his being is, and to the xtent that *it is* and nothing else is, it is one and simply that. And nothing else can e said to participate in it in any proportional way.

The theory with regard to the transcendentals, however, holds that there is convertibility between being, one, true and good in different beings. To the xtent that a thing is, it is one, true and good. But for Parmenides there is no versity of being. There is being in the sense of one being, a being, which is. here is nothing else. The transcendental theory states that there is a convertibility etween being and the transcendentals, although the transcendentals may add a new onnotation. In the case of oneness, for example, the notion of indivision is added, ut this does not add anything to the *being* of being. In n. 560 of *In IV Metaph.*, . Thomas says:

> The one, which is convertible with being, designates being itself, and adds the notion of indivision, which, since it is a negation or privation, does not state any nature added to being. And thus, in no way does it differ from being in reality but only conceptually. For negation or privation are not beings of nature but of reason....

In proving that being is one in the argument of Parmenides as given here i the *In Physic.*, Parmenides first proves that there is nothing but being, since, there were anything else, it would be nothing:

Other than being = non-being
Non-being = nothing
Ergo, other than being = nothing.

And then by replacing the term being by the term one, Parmenides prove that there is only one being. What he is actually doing is confusing *a* being wit being and *a* one with one. And since being is one and not two or many, there i nothing else that exists and, consequently, there is nothing that can move it.

57. Parmenides took the noun "being" and considered it as the participle c *be-ing*. According to Parmenides, wherever one speaks of the being of anything no matter what the thing may be, one always speaks of the same thing. One thin *is* in the same way as any other thing is. He had no notion of the meaning c analogy. He takes the particular being and assumes that we always say and mea the same thing in speaking of "to be." He is not aware of the fact that being i said in a variety of ways and that for this reason we must take care to show in wh sense we say and mean that whatever is "is." For Parmenides, to speak c anything and to say that it exists is the same thing. For example, when we spea of trotting horses at a race, we may say "they are running." This implies that eac is and is running. This example would be comparable to the meaning of Pa menides, who, if he said that all things are, would imply that each is, and he woul mean that each is a thing in exactly the same way as any other thing is.

8. This is one of the instances which, Commentators agree, indicates that .ristotle recognized the *fact* of analogy, although not the *theory* of analogy. Plato, owever, neither stated nor expounded the theory of analogy but it is implied in the 'ay he speaks, e.g., in the *Parmenides*.

9. One might well compare this with the playing on the term "one" so eautifully done in the *Parmenides* of Plato. The ambiguities lurking in the ıeanings of the term "one" are endless. The fluctuating connotations of the xpression "If the One is" could well be the subject of an entire volume.

0. If we replace the term "being" in the argument of Parmenides by the term white," and if we say that "whatever is not white is nothing," and then go on to ay that white is one only, Aristotle says this is ridiculous for three reasons. White annot be one:

1. *As the continuum is one*: Even if everything were white, this would not prevent the existence of many white things, e.g., many white particles in one piece of chalk, or even many pieces of white chalk. And so, the argument, as applied to being, does not hold.
2. *As one by continuation*: To be one in this way is to express the quality of unity in the thing that is one as a continuum. All magnitude is divisible but inasmuch as it is divisible, it is in many in potency but one in act. It is one by continuation.
   a. There are various ways of being together, i.e., continuity, contiguity or colligation, or nearness to one another, as St. Thomas says in the *In IV Physic.*, n. 685.
   b. If everything white were one by continuation, i.e., one single continuous thing or one solid mass of being, this would denature the meaning of a continuum, since a continuum is divisible into an infinite number of parts,

as St. Thomas said in n. 3 of the third lesson of this book.

c. The position of Parmenides would be similar to that of one who supposed the whole universe to be one organism. If this were the case, in some sense the whole of being would be one individual being.

3. *As one in meaning*: This would mean that white and the thing that is white would be one. Man and white would be one conceptually. This would be ridiculous. And so, if this argument were applied to the argument about being, the latter argument would be rendered absurd, for man and being, star and being, would all be one in meaning. And they are not. A subject and the accident predicated of it are not the same conceptually, although they may be one in subject. This would imply that being has one univocal meaning and only that.

61. The being of Parmenides would seem to signify what is one as one indivisible mass.

In less. 7 of the *In V Metaph.*, St. Thomas considers three ways in which things are called one "incidentally" and five ways in which things are called one "strictly." (We are translating "per accidens" by "incidentally" and "per se" by "strictly," realizing that these are not adequate translations.) St. Thomas considered that something had unity if it had the unity of indivisibility, when he says in n. 865:

> Aristotle says that those things are 'completely,' i.e., perfectly and most one, of which the concept apprehending their quiddity is completely indivisible, as simple things which are not composed of material and formal principles. Hence, the concept that apprehends their quiddity does not comprehend them as if composing their definition from various principles, but rather by way of negation, as a point that has no part, or even by way of

> disposition to composites, as if it is said that unity is the principle of number.
>
> And, because the concept itself of such things is indivisible, and since those things that are divided in any way can be known separately, it follows, therefore, that such things are inseparable not only in time, but also in place and meaning. And for this reason they are most one, especially that which is indivisible in the genus of substance; for what is divisible in the genus of accident, although in itself it is not composed, is still a composite for another, i.e., for the subject in which it is. But indivisible substance is neither composed in itself nor is it a composite for another. Or, the word substance can be in the ablative case. And then, the sense is that, although some things are called one because they are indivisible in place or time or meaning, yet, among these things, those are called most one which are not divided in substance....

First of all, Parmenides, in saying that "being is one," did not say in what way being was one. If it is taken as one in the way that the indivisible is one, it can be one according to the indivisibility of

1. Time
2. Place
3. Meaning
4. Substance
5. Concept of substance.

It would seem that the fourth one would be the way in which Parmenides meant "one," for the others do not seem at all proper to the tenor of his argument. If the Parmenidean being were an indivisible one as an indivisible uncomposed substance, it would be impossible to account for the obvious multiplicity in the universe except *by appearance*, as Parmenides seems to admit in his "Poem," but not *in reality*.

62. White and that of which white can be predicated are not identical in meaning A horse is white. Horse and white are not the same in meaning. A horse ca jump. White cannot jump, but a white horse can jump. Although white and hors may be one in subject, there are two significations in saying *white horse*, viz. *white* and *horse*. White cannot be separated from white horse entitatively or i subject, but it can be separated from white horse in thought or meaning. White ca be separated from horse both in reality and in meaning, for white need not b predicated of every horse, since every horse is not white.

The consideration here, of course, concerns contingent accidents which ca be separated from their subjects in reality, but it also holds true for prope accidents which, although they cannot be separated from their subjects entitatively can be separated and distinct in meaning. Parmenides did not establish o recognize the distinction between subject and accident. His is primarily an erro of predication, since he made all predication to be identification. According t him, white would be being in the same way as horse.

63. To say of something that it is substance is to say that *it truly is*, but it is no to say of it that *it is what truly is*. The latter would imply that anything other tha this would not truly be. Suppose we say of man that he is a substance. In s doing, we are saying that he truly is but we are not saying that only man truly is

Substance is being but not all being is substance, and so, Parmenides argument would result in the absurd equating of being with substance. Being i much more universal in its comprehension than is substance. Substance is, so t speak, a contraction of being in that being is divided in several ways, and amon these, into substance and accident.

64. Here it is shown that if in the statement: "Whatever is other than being is non-being," the term "being" were to signify "accident" so that the latter could replace the former, and the statement would read: "Whatever is other than accident is non-accident," this would lead to absurdity. Since accidents are said of subjects, there would be no subjects of which accidents might be predicated, for whatever is other than accident would be non-accident, which, according to this argument, would be non-being. But subject is other than accident, and so, subject would be non-being.

More specifically, if we take sweet as an example of accident, and then say that whatever is other than sweet, which, according to this argument, would be all that is, is non-sweet, there would be nothing of which sweet might be predicated, because only sweet would exist. But, according to the philosophy of St. Thomas, sweet is an accident and it is the nature of an accident that it be predicated of something. It would then, according to the Parmenidean understanding, have to be predicated of non-being. And so, accident or sweet would be predicated of non-being. This whole system would be ridiculous, since it is the nature of contradictories, i.e., being and non-being, the fundamental contradictories, that they not be predicated of one another, as is shown in Appendix note 65.

It is true to say that whatever is other than a being is the non-being of that being, but it is not non-being absolutely. Thus, it is not always true to say that whatever is other than being is non-being. For example, accident is a being, and therefore, to say that something is other than accident is to say it is other than a being, viz., accident, and not other than being, in the sense that it is non-being.

Sweet is an accident. Whatever is other than sweet is other than an accident, viz., this accident or sweet, but it is not other than accident in the sense that it is non-accident absolutely, which, according to the argument of Parmenides, would be non-being absolutely.

65. Since what is said in *In IV Metaph.* is pertinent not only for the resolution of the argument of Parmenides, with which we are now concerned, but also for the statements of Heraclitus and of Anaxagoras with whom, among others, we are concerned here in the first book of the *In Physic.*, a summary of the treatment of this basic principle of non-contradiction is helpful for a more complete understanding and criticism of the problems involved.

(Here the Pirotta edition refers to less. 16 of the *In IV Metaph.* This would seem not to be the correct reference, since this lesson deals with the fact that there is no medium between contradictories rather than with the fact of the impossibility of the mutual predication of contradictories. It would seem that lessons seven through fifteen would be more appropriate for the proof that contradictories cannot be predicated of each other.)

In *In IV Metaph.*, less. 6, St. Thomas says that metaphysics, because it is the science that has the greatest certitude, should consider the most certain principles of science. He gives three conditions for the most certain principle: first *one cannot err with regard to it*; secondly, it is *unconditional* and is subjected to nothing; thirdly, it is *known naturally* and not through demonstration. In n. 600, he states that this principle is the principle of non-contradiction, viz., that the same attribute cannot be in and not in the same thing at the same time and in the same respect.

The remainder of the sixth lesson proves that the three conditions mentioned are fulfilled in the principle stated. This principle, the first principle of demonstration, is considered in metaphysics, but is used in every science, since the proper operation of a science is to know and demonstrate what can be said of its subject matter. In the particular sciences, there is always some contraction of the being that is treated in metaphysics. Each science, with the exception of metaphysics, is limited to a particular kind of being, but the medium of proof, viz., demonstration, is the same in the particular sciences as in metaphysics. And so, the first principle of demonstration, i.e., the principle of non-contradiction, is the same in all sciences. There is a veritable appropriation of the common principles of metaphysics by the particular sciences which suppose and use metaphysics. There is no certitude in any science except by resolution to first principles. And, since the certitude of a science depends upon the certitude of principles, it is well for us to see why the principle of contradiction, which has been declared the most certain principle, is true. Obviously, this principle is indemonstrable or it would not be first but dependent, and some kind of regression would be necessary to prove it.

St. Thomas, however, takes great care to show its truth by means other than demonstration. Although the principle cannot be demonstrated, first principles demand that the intellect see with necessity the falsity of the opposite of first principles. To have absolute certitude we must totally reject the opposites of the principles involved. Otherwise, we cannot speak of absolute certitude. Lessons seven through nine offer seven general arguments against those who deny the principle of non-contradiction. As his hypothesis in proceeding against these, St. Thomas assumes that a noun means something determinate and distinct from its contradictory. The intellect must fix itself on something determinate. It cannot fix itself on nothing. If nothing is known, nothing is signified. If the intellect does not signify something one, there is no signification, because there is no knowledge. There is signification in the measure that there is intellection. If words signify indefinites, there will be no argument. The "infinite" or indefinite signifies "something," but in a negative way in that it does not signify a determinate thing but determinately signifies all that is not man, for example.

The noun signifies something determinate. The noun man does not signify not being man. Words expressing the nature or essence of man do not signify not being man. The noun man and the noun not man do not, therefore, signify the same thing. In fact, they signify contradictories and these cannot be predicated of each other.

The first of the seven arguments that are given to prove that contradictories may not be predicated of each other is given in less. 7, nn. 613–624. In this proof we must keep in mind that St. Thomas is proceeding elenctically from the opponent's concession that the noun signifies something.

After saying that man signifies something and that not man signifies something, St. Thomas says in n. 619:

> He proves the third point, viz., that man and not man do not signify the same thing, and he proves it by the following argument. Man signifies this, viz., to be man and the nature of man (quod quid est homo). But not man signifies not being man and the nature of non-man (quod quid est non homo). If, therefore, man and not man do not signify something diverse, then, what is being for man will not be different from what is non-being for man, or not being man. And thus, one of them will be predicated of the other. And they will also be one in meaning; for, since we say that some things signify one, we know that they signify one meaning, e.g., coat and garment. If, therefore, being man and not being man are one in this way, viz., in meaning, that which will signify what it is to be man and what it is not to be man will be one and the same. But it has been given or shown that there is a different word that signifies each; for it has been shown that this noun man signifies man, and it does not signify not being man. It is evident, therefore, that being man and not being man are not one in

> meaning. And thus, what was proposed, viz., that man and not man signify different things, is evident.

Being man and not being man will always be distinct things in reality, and so, they will not be predicated the one of the other. St. Thomas makes this clearer in n. 622, where he says:

> Here he proves something that he had supposed. Moreover, to prove that this name man does not signify what not being man is, he assumed that what being man is and what not being man is are different, although they may be verified of the same thing. And he intends to prove this by the following argument: To be man and not to be man are more opposed than man and white. But man and white are different in meaning, although they are the same in subject. To be man and not to be man are, therefore, different in meaning. He proves the minor in this way: For, if all things which are said of the same thing are one in meaning as if signified by one name, it follows that all things are one, as was said and explained above. If, therefore, this does not happen, that which was said, viz., that to be man and not to be man are different, will happen. And the last foregoing conclusion, viz., that man is a biped animal and that it is impossible that this is not a biped animal, consequently follows.

A biped animal cannot be what is not a biped animal. And what is not a biped animal cannot be a biped animal. Nor is being non-being, nor non-being being. Contradictories may not be predicated of each other.

In less. 10 of the *In IV Metaph.*, St. Thomas argues more specifically against those who, since they did not hold the principle of non-contradiction, said that contradictories could be simultaneously true. In n. 665, he says:

> Aristotle says, therefore, that the opinion with regard to this, viz., that contradiction may be verified simultaneously,

> came to some by way of doubt from sensibles in which generation, corruption and motion are apparent; for it seemed that contraries came to be from some one thing, as air which is warm and earth which is cold came to be from water. But everything that comes to be comes to be from something that is beforehand; for it so happens that what is not does not come to be, since nothing comes to be from nothing. A thing would, therefore, have to have a contradiction in it at the same time, since, if the hot and the cold come to be from one and the same, then, it becomes *both* warm and not warm.

In n. 666, he shows how Anaxagoras and Democritus fell into error in believing that contraries came to be out of some one thing:

> Because of this reasoning Anaxagoras said that all things are mixed in all things; for from the fact that he saw that anything whatever came to be from anything whatever, he thought that nothing could come to be from another unless it was in that other beforehand. And Democritus also seemed to go along with this reasoning; for he stated that the full and the void were joined in any part of the body whatsoever. These, indeed, are related as being and non-being; for the full is as being, but the void is as non-being.

St. Thomas answers the problem by saying that these philosophers were partly correct and partly incorrect. In n. 667, he says:

> Being is said in two ways: being in act and being in potency. Since they say, therefore, that being does not come to be from non-being, what they say is in some way true and in some way not; for being comes to be from what is in act non-being and from what is in potency being. Hence, in some way the same thing can be being and non-being at the same time, and in another way, it cannot; for it so happens that the same thing is potentially contraries, but not 'perfectly,' i.e., in act. It is in this way, indeed, that the lukewarm is in potency hot and cold, but in act it is neither.

St. Thomas devotes the eleventh through the fifteenth lessons of the *In IV Metaph.* to showing the method and error of those who thought that truth consisted in what they saw, i.e., in sensible appearances. In this consideration, St. Thomas speaks of the errors of Anaxagoras, Empedocles, Parmenides, Heraclitus and others. This view of St. Thomas seems a bit astigmatic, for, although Heraclitus repeatedly spoke of changing appearances and the appearance of change in the sensible world, Parmenides at least seemed to strive for the whatness behind the seeming but not real multiplicity in the universe. It would not seem that the being of Parmenides is a being arrived at by one who stressed sensible appearances. Nor did Anaxagoras arrive at his mixture of infinitesimal parts, each of which contains the other, through conclusions from sensible phenomena only. Moreover, the cause of separating and mixing as posited by Empedocles, viz., love and strife, was not a sensibly apparent cause in the formation of the universe.

In speaking critically of the foregoing philosophers, St. Thomas says that, contrary to what these men concluded from appearances, neither contradictories nor contraries can be in the same thing at the same time and in the same respect. The sixteenth lesson is devoted to showing the error of those who, since they did not hold the principle of non-contradiction, said that all things were simultaneously true and false. In n. 736, St. Thomas says:

> He disputes against certain positions which are consequent upon the foregoing: and first against those who destroy the principles of Logic, and secondly, against those who destroy the principles of Physics.... For the first philosopher must dispute against those denying the principles of particular sciences, because all principles are founded upon this principle, viz., that affirmation and negation are not simultaneously true, and that there is no medium between them. But those are most proper to this science, since they follow upon the notion of being which

> is the first subject of this philosophy. But the true and the false pertain properly to the consideration of the logician, for they follow upon being of reason which the logician considers, for the true and the false are in the mind.... The error with regard to the true and the false follows upon the error with regard to being and non-being; for the true and the false are defined through being and non-being.... For there is truth when it is said that what is is or that what is not is not. But that is false which is the reverse.... And, therefore, if the errors with regard to being and non-being are destroyed, so also will the errors with regard to the true and false...be destroyed.

Affirmation and negation are rooted in being and non-being as their foundation and so the impossibility of the simultaneous truth of contradictories is rooted in the impossibility of the same thing's being simultaneously itself and not itself. If we are to say with St. Thomas that the whole intelligible or intentional order is founded upon the real order, the problem of the affirmation and negation of contradictories depends upon the co-existence or non-co-existence of contradictories in the real order. Affirmation and negation are signs of the act of the intellect that composes or divides in an enunciation. Affirmation and negation are consecutive to composition and division, for the intellect, when it does so truly, composes or divides according to what is or what is not. Just as there is a clear distinction between what is and what is not, so also is there between what is composed and what is divided and between affirmation and negation. Yet, there is a close relationship of being and non-being to composition and division, and of composition and division to affirmation and negation. Being and non-being are respectively the cause of affirmation and negation. When we say this, however, we must not mean that this cause is despotic, for there always remains a radical indetermination of the intellect to what is true and to what is false. The intellect, if it is to arrive at truth, must conform to what is, for it is the intellect that composes what is identical and divides what is distinct. We must remember that it is also the intellect which errs, and only the intellect can decide between the two. It is the intellect functioning properly, which, by conforming with what is, gives rise to affirmation and negation and thus to truth or falsity, dependent upon whether or not the intellect has properly conformed. We cannot understand affirmation and negation without considering conformity to reality. It is this conformity with reality that gives truth or falsity. We are here, of course, concerned with speculative truth and speculative error and not with the practical order.

In short, the principle that says that it is impossible to affirm and deny the same thing at the same time depends upon the principle which holds that it is impossible that being and non-being be in the same thing at the same time. And between these two principles there is a relation of dependence, as we have said before. The former is dependent upon the latter. The former principle, then, is a logical principle based upon the metaphysical principle. Contradictories, therefore, cannot be predicated of each other in the logical order, since they cannot be at the same time and in the same respect in the order of objective reality.

66. Here it is shown that in the statement: "Whatever is other than being is non-being," we cannot understand that being means substance and then formulate the proposition: "Whatever is other than substance is non-substance," in the sense that non-substance is made equivalent to non-being. If substance is being, and whatever is not substance is non-substance or non-being, since accidents are not substances, accidents will be non-being. If it is stated that substance is white, this would mean that the accident white, which, according to the above reasoning, is non-being, would be predicated of being, i.e., of substance. In other words, to say that

substance is white would be to say that it is what is not, since, according to the foregoing, the accident white would be non-being. And thus, we would be predicating non-being of being.

For example, if we take wood in the place of substance, whatever is other than wood, is not wood, i.e., it is not this substance, which is wood. Whatever is other than this substance is certainly not this substance. The fact that not wood is other than wood or not being wood does not mean that it is non-being absolutely speaking and as such, as Parmenides would have it.

Hence, just as in the case of substituting accident for being, we found that there would result the absurd consequence of saying of being that it is non-being, i.e., that accident (being) is substance (non-being), so also here in the case of substituting substance for being, saying of non-being (accident) that it is being (substance) would result. Being and non-being, the basic contradictories, cannot be predicated of one another, as was said in Appendix note 61 and substantiated in Appendix note 65.

67. Since being is said in a variety of ways, the way in which we mean being must always be considered. If being were univocal, then, non-being would be nothing. This is the Parmenidean problem of being. If being is univocal, what can a negative proposition mean? John is not white would mean that John is nothing. It is not necessary that not being be nothing because not being can be being other.

Being has as many meanings as there are kinds of beings. Every kind of being in the universe is some communication of the act of being. There is everywhere a participation of all beings in being or existence and there is also a proportion of essence to existence. Being is a concept neither of univocity nor of equivocity. The analogy of being is a metaphysical analogy involving a proportion between potency and act. It is potency and act that makes real analogy possible. (Cf. W. Norris Clarke, S.J., "The Limitation of Act by Potency," *op. cit.*) Essence is as the potency and existence is as the act in the proportion. Essence is always proportioned to existence. It is, so to speak, the mode and channel of existence or a "structured existence."

Existence is a basic perfection and there is participation of this common perfection diversely in many beings. Each existence *is* in a different way from any other. Thus, the analogy of proportion and that of proper proportionality are combined in this theory of participation. Here, however, Aristotle is referring exclusively to an analogy of attribution, i.e., to the relation of substance in the analogy of being. It seems that St. Thomas is also using only the analogy of attribution here, for this is all that is necessary in this problem.

In recent years, many Thomist philosophers have emphasized the dynamism and impact of the theory of the analogy of attribution combined with that of proportionality. Analogy makes possible a way of thinking where precisely the like contains precisely the unlike. Being has a proportional nature. A proportion is a direct relation of one thing to another, whether it be a relation of excess or of defect. And so, the proportion that will be universal in being will be the proportion between essence and existence or between potency and act. We can never break through to the nature of the pure being that is pure existence. We never penetrate beyond essence to pure existence. Existence in things is limited by essence, for things have existence in proportion to their essence.

The idea of being includes the difference between any two existences. It includes the like and the unlike in both an analogy of attribution and of proportionality. Being is immanent in every particular diversity. It is likeness in which there is unlikeness. Being will always be proportioned and the proportions that are possible are endless.

Analogy is an instrument for solving the dialectic of contraries which troubled Parmenides and other ancient philosophers. Analogy is, for those who follow St. Thomas and for others, too, the last of a series of attempts to solve the relationship of the contraries to each other. The history of philosophy, especially of ancient philosophy, has been a search for the interpenetration of being and non-being, the like and the unlike, and the one and the many. Parmenides tackled and toiled with the problem of being and non-being. We agree in part with him that it is impossible to consider anything but being itself that would unite beings, for we, too, believe that there is an all-pervading unity in the ensemble of things and that this unity is that of being. But we believe that limited beings have a manner of non-being, e.g., man is non-apple. It is this that stymied Parmenides. We believe that the limited modes of being are being but only according to a proportioned measure of being. Parmenides did not see this, and so, he said that all was a being. According to the theory of analogy, the manner of non-being that is in beings must not be considered apart from the world of being. The theory of analogy, which is an answer to the problem of being and non-being is an answer to the problem of Parmenides that has interested philosophers for centuries.

In the dialogue *Parmenides*, we find that there are eight descriptions of the being of Parmenides. These various descriptions are not explicitly outlined in the dialogue, but offer a way in which one might schematize the various aspects of the problem of being in the Parmenides.

1. It is.
2. It is uncreated and indestructible.
3. It is complete, not incomplete.

4. It is immovable and without temporal end, nor will it be nor was it, for then it is not.
5. It is indivisible. There is a definite intention of keeping out the interpenetration of non-being here.
6. It is wholly continuous and without origin. It continues into things so that it reduces and does not diversify them.
7. It is immovable and not infinite, for then it would stand in need of everything. Being is complete on every side, like the mass of a rounded sphere equally poised from the center in every direction and reaching out equally. This is a corporeous simile of being as a continuum.
8. It is not anything empty, for anything empty is nothing.

And our answer to the problem to which the above kind of being gives rise is the notion of participated being as presented in the theory of analogy. There is not just one meaning of being, but there are many things that are and each of these *is* in a different way. Being is said in as many ways as there are proportions of essence to existence. A substance is a kind of being and an accident is a kind of being. The very principle of unity, analogous unity, is the proportion between essence and existence and it is the very principle of unity that determines diversity. Being is found proportionately in different beings, as we have said, and it is immanent in every particular diversity.

68. If being had only one meaning, it could not be an "accident with a subject," for there would be no way of distinguishing the accident from the subject of which it would be predicated. Besides, the very consideration of an accident and a subject implies a duality. Subject and accident are different kinds of being.

Moreover, being, if it is to have one meaning only, could not be quantified, since this, by its very nature, would imply divisibility into parts, no one of which is identical either with the whole or with another part. Thus, being, if it were quantified, would imply a diversity. This is why St. Thomas says that being, if it is to have one meaning, could not be a corporeal substance. Corporeity implies a multiplicity either from the viewpoint of subject and accidents, or from the possibility of division into distinct parts.

69. We have been told by St. Thomas in *In I Metaph.*, less. 9, n. 139, that being is not a genus and that it does not have one nature only, as that of a genus or of a species. In Bk. III, less. 8, n. 433 of this same work, he proves that neither *being* nor *one* can be a genus:

> Since the difference added to a genus constitutes a species, therefore, it will not be possible to predicate of the difference either species without the genus or genus without the species. But that the species cannot be predicated of the difference is obvious for two reasons: first, indeed, because the difference has a wider extension than the species, as Porphyry said. The second reason is that, since the difference is stated in the definition of the species, the species cannot be predicated strictly (per se) of the difference, unless the difference were taken as the subject of the species, as number which is stated in the definition of even, is the subject of even.
>
> But this is not the case, but rather the difference is, so to speak, a form of the species. The species, therefore, cannot be predicated of the difference, unless per chance accidentally.
>
> Likewise, the genus taken strictly (per se) cannot be predicated of the difference in the mode of predication strictly speaking (per se); for the genus is not stated in the definition of the difference, because the difference does not share in the genus, as is said in the fourth book of the *Topics*. Nor is the difference stated in the definition of the genus. The genus, therefore, is predicated of the difference in no mode of strict (per se) predication. It is, however, predicated of what 'has a difference,' i.e., of species wherein the difference is actual. And, therefore, he says that the species may not be predicated

> of the proper differences of the genus; nor is the genus without the species so predicated, because the genus is predicated of the differences according as they are in the species. But there can be no difference of which being and one are not predicated, since any difference whatsoever of any genus whatsoever is being and is one; otherwise, it would be impossible for it to constitute some one species of being. It is impossible, therefore, that one and being are genera.

Now that we have shown that, according to St. Thomas, being is not a genus, we now refer to *In V Metaph.*, less. 9, nn. 885 ff. to show that being has a multiplicity of meanings. In summarizing what St. Thomas says, we find the following divisions of being:

I. Being is divided into
   - A. Accidental (per accidents) being and being strictly speaking (per se).
     1. There are three modes of accidental being.
        - a. Accident is predicated of accident, e.g., the just is musical.
        - b. Accident is predicated of a subject, e.g., man is musical.
        - c. Subject is predicated of accident, e.g., the musical is man.
     2. The divisions of being strictly speaking included division of being into the ten predicaments. These are substance, quantity, quality, relation, habitus, time, place, position, action and passion. Nine are in the genus of accident. These kinds of being are extramental and he calls them "perfect being." Upon these kinds of being strictly speaking are based the modes of predication. In *In V Metaph.*, less. 9, nn. 889–890, St. Thomas says:

> It should be known that being in this mode cannot be contracted to anything determinate, as a genus is contracted to the species through the differences; for the difference, since it

does not share in the genus, is outside the essence of the genus. But outside the genus of being there can be nothing which, by adding it to being, constitutes some species of being; for what is outside being is nothing, and cannot be a difference. Hence, in the third book of this work the Philosopher has proved that being cannot be a genus.

Hence, being must be contracted to different genera according to a diverse mode of predication which follows upon a diverse mode of being (essendi)....

B. Mental being or being in the mind. This is logical being, the being of propositions, the true and the false established through negation and affirmation, and composition in propositions; and only in compositions or division is such being in act.

C. Being in act and being in potency. Both extramental and mental being may be considered as actual or potential. One might wonder how it is that mental being can be so considered. In their very notion, mental beings may mean something in act or something in potency. Substance, for example, is actually a genus, and as such, it is potentially a species.

It, therefore, seems quite obvious that being is indefinable, since it is neithe a genus nor a species, and since it has a variety of meanings, none of whic includes the complete nature of being. Now that we have seen that being as suc is not definable, it is not difficult to show that it is not a definable substance. I fact, there is no such extramental reality as substance but only substances. Ther are many substances; there is not just one kind of substance. In the tenth lesson o the *In V Metaph.*, St. Thomas says that there are four kinds of substance. It ma be taken to mean:

1. *Particular substances*, which he calls first substance, e.g., this stone or this flesh. These substances are not predicated

of a subject, but other things are predicated of them. This division of substance is explained in Appendix note 39.

2. *The cause of being (essendi) in first substances*, e.g., the soul is the substance of animal.
3. *Parts of a thing which are the terminations of that thing*, e.g., a line is a part and termination of a surface. He says this is not a real kind of substance, although the Platonists and Pythagoreans thought so.
4. *A thing's essence or nature which is expressed in the definition*. In this way a thing's substance is its genus and species which constitute the definition.

And in n. 903 of this same lesson, St. Thomas reduces the above four modes of substance to two, viz., to first or particular substance and to second or universal substance. First substance, e.g., this man, is "this something," which is self-subsistent, incommunicable to others and not common to many. If we are to agree with St. Thomas, second substance, e.g., man, is common to many. It can exist only through the particular. Yet, as actually existing in the latter, it is no longer second substance but first substance.

St. Thomas lists three differences between particular and universal substance:

1. Particular substance is not predicated of any inferior, as is universal substance.
2. Universal substance only subsists in the nature of the singular which is self-subsistent.
3. Universal substance is in many things, but singular substance is not, but is separable and distinct from all others.

For St. Thomas universal substance is second substance, and he cites man, not this man, as an example. It seems difficult to hold, as St. Thomas does, that there is such a being as second substance. There is, however, a sense in which substance is universal in his understanding of first substance, for all first substances are universally substance. Just as being is said analogously, so also is substance.

There are many kinds of being; there is not just being. And there are many kind of substance; there is not just substance.

Another division of substance is expressed in the beginning of the *De Ent et Essentia*, chap. 1, *op. cit.*, pp. 12–13, where St. Thomas speaks of simple an composite substances:

> Since being is predicated absolutely and primarily of substances, and secondarily and, so to speak, qualifiedly of accidents, thus, essence is properly and truly in substances but only in some respect and qualifiedly in accidents.
>
> Some substances, moreover, are simple and others are composite. And yet, in both cases there is essence but it is in simple substances in a truer and more noble way, inasmuch as they also have being in a more noble manner. They are also the cause of the substances that are composite. At least, the first and simple substance, which is God, is such a cause.

Before launching upon his study of simple and composite substances here i the *De Ente et Essentia*, St. Thomas reminds us that in accordance with our wa of knowledge we should start with the study of composite substances, since thes are less obscure than are simple substances.

70. An accident is said "per se" of something, when its definition must includ that of which it is predicated.

In the *De Ente et Essentia*, chap. 7, *op. cit.*, pp. 53–55, St. Thomas says:

> It should also be known that genus, species, and difference are taken in a different way in the case of accidents than they are in the case of substances. The reason for this is that in the case of substances, a "per se" unity is brought about from the substantial form and the matter, and some one nature results from the union of these. This nature is properly placed in the predicament of substance. In the case of substances, therefore,

the concrete names which signify the composite are properly said to be in a genus, as species or as genera, e.g., man or animal. But neither the form nor the matter is in a predicament in this way, except by reduction, as principles are said to be in a genus.

But a "per se" unity does not result from an accident and a subject. Hence, some nature to which the notion of genus or of species could be attributed does not result from their union. Hence, the accidental names, said concretely, are not stated in the predicament as species or as genera, e.g., white or musical, unless by reduction, but only as they are signified in the abstract, e.g., whiteness or music. And because accidents are not composed of matter and form, therefore, the genus in them cannot be taken from the matter and the difference from the form, as in the case of composite substances.

But the first genus must be taken from the very mode of being (essendi), inasmuch as being is predicated of the ten genera in different ways according to the order of priority and posteriority. Thus, it is called quantity from the fact that it is the measure of substance, and it is called quality from the fact that it is a disposition of substance, and so, too, with regard to the other accidents....

But the differences in accidents are taken from the diversity of the principles from which they are caused. And because proper attributes are caused from proper principles of a subject, therefore, in the definition of those things, the subject is stated instead of the difference, when they are defined in the abstract. In this latter kind of definition, the proper attributes properly belong in the genus, as it is said that snub-nosed-ness is in the curvedness of the nose. But the converse would be true if their definition were taken as they are said concretely. For in this way the subject would be stated as genus in their definition, because then they would be defined in the manner of the definition of composite substances in which the generic notion is taken from the matter, as we say that a snub nose is a curved nose....

71. With regard to the first postulate, St. Thomas says that if animal and bipe are parts of the definition of man, they cannot be contingent, i.e., separabl accidents of man, because then they would not be essential parts of the definitio but parts that at will might or might not be in the definition.

Now there are also accidents that are not separable, viz., accidents tha require the subject in their definition, as does the definition of pug-nosed. Nos must be stated in the definition of pug-nosed, which, though it is a "per se accident of nose, is still an accident. It is not of the essence of a nose that it b curved, but it is an accident that can be predicated only of this kind of subject. I is peculiar to snubness that it be predicated of nose. So also is white a "per se accident of surface, since surface enters into the definition of white.

According to the second postulate the definition of the thing defined canno be stated in the definition of a part of the thing defined, i.e., the definition of th whole cannot be stated in the definition of a part. Since nose is stated in th definition of pug-nosed, pug-nosed will not be in the definition of nose. Moreover since surface is in the definition of white, white cannot be in the definition o surface. And since biped is stated in the definition of man, man cannot be state in the definition of biped. If this were not so, man would be known both befor and after his essential parts are known, i.e., before and after definition.

72. In the *Topica*, Bk. VI, chap. 14, Aristotle speaks of the formulation o definitions. And in the course of this discussion, he says that we must be sure tha the definition is all-inclusive.

And he says that if we are not sure that a given definition is an adequat definition of the thing defined, and if we are not able to deal with the whole, we must begin by checking at least some part of the definition to see if it is correct.

Obviously, if the part is incorrect, the definition is vitiated. If the part which is known *in itself* before it is known *as a part* of the definition is not correct, neither will that which is later composed of the part be correct. With regard to the definition, Aristotle says that the part is known in itself before it is known as a part of the whole and, therefore, if this part is unknown or inadequate, the definition in which it is stated will be inadequate.

In other words, if the part which is known as a part of the definition is incorrect, the whole definition is incorrect, because we know the whole through the parts. We define composites through their parts, and so, it is obvious, since we define things as we know them, and we know them as they are, that we must seek and strive to know them as they are. This is a part of the investigation of truth.

73. The specific difference is part of a definition, not part of a being. It is never an accident. In a definition of a thing, the specific difference is that part of a definition which corresponds to the form, e.g., rational in the definition of man which states that he is a rational animal. Now, rationality is not an accident of man. It is a part of his very quiddity, which is essence looked upon as corresponding to definition.

In the third chapter of the *De Ente et Essentia* St. Thomas gives a good explanation of genus and specific difference which together constitute a definition. On pp. 18–21, *op. cit.*, we read:

> So, therefore, it is clear that the essence of man and the essence of Socrates only differ according to designated and non-designated: hence, the Commentator (Averroes) says in the seventh book of the *Metaphysics* that 'Socrates is nothing else but animality and rationality, which are his essence....' In this way, too, the generic essence and the specific essence differ in

> that they are designated or non-designated, although there is a different mode of designation for each of them, because the designation of the individual with respect to the species is through matter determined by dimensions; but the designation of the species with respect to the genus is through the constitutive difference, which is taken from the form of the thing.
>
> But this determination or designation, which is found in the species with respect to the genus is not through anything existing in the specific essence which is not in any way in the generic essence; rather, whatever is in the species is also in the genus as undetermined. For, if animal is not the whole that man is, but a part of it, it would not be predicated of him, since no integral part is predicated of its whole.

St. Thomas goes on to give the example of body to clarify the above example. He says:

> We can see how this is so if we consider how body differs as stated as *part* of animal and as stated as a *genus*, for it cannot be a genus in the same way that it is an integral part. This word body, therefore, can be taken in a variety of ways; for body, inasmuch as it is in the genus substance is said of that which is of such a nature that three dimensions can be designated in it. These three designated dimensions, however, are body, which is in the genus quantity....
>
> On the other hand, what the difference is is determined by the form taken in a determined way, without determinate matter's being in the first understanding of it, as is evident when we say animate, viz., that which has a principle of life, for what it is, whether body or something else, is not determined...and, therefore, as the Philosopher says in the third book of the *Metaphysics*, and in the fourth book of the *Topics*, the genus is not predicated of the difference, strictly speaking, except, incidentally, as a subject is predicated of an attribute. But the definition or the species includes both, i.e., the determinate matter, which the name genus designates, and the determined form, which the name difference designates.

From the above lengthy citation, it should be obvious that what belongs to the specific difference and to the genus belongs to the essence and, therefore, cannot be an accident. "Biped" belongs to the specific difference in the definition of man. It cannot, therefore, be considered as a separate accident. And since man is not stated in the definition of biped, any more than man is stated in the definition of animal, it cannot be an inseparable accident either.

Rather, biped, like risibility, is related to the form or specific difference in a given definition of man. The difference, of course, is part of the definition, *not* part of the being. It is well, too, to mention here that the definition is necessarily an arbitrary way of stating something. *A* way of defining man is to say "biped animal," but both Aristotle and St. Thomas recognized later that the true definition of man is "rational animal." To return to the statement of the relationship of "biped" and "risible" to the form, St. Thomas says in the seventh chapter of the *De Ente et Essentia*, *op. cit.*, p. 53:

> And because each thing is individuated by matter, and is placed in a genus or species by its form, therefore, the accidents that follow upon matter are accidents of the individual, and according to these accidents, individuals of the same species differ from one another. But the accidents that follow upon form are proper attributes either of the genus or of the species. And so, they are found in all things that share in the nature of the genus or of the species, as risible follows upon the form in the case of man, because laughter occurs as a result of some knowledge on the part of the soul of man.

74. Here it is not meant that man actually is "*what* truly is," i.e., substance, but that man is something which truly is or a-that-which-truly-is. Man is not substance as such or considered absolutely, but he is *a* substance.

75. If there were just one being, one part of a definition would signify the whol e.g., either biped or animal separately, or together. Or the whole, e.g., ma would signify either biped or animal separately or together. There would be n possibility for grounds of differentiation in physical reality or in definition betwee the whole and the parts. Any animal would be man, and vice versa. Thi univocity is, of course, illogical. Parts of definitions are not identical with wha the total definition signifies any more than it is necessary for a part to be the sam as the whole in the real order. If there were just one being, man would have n parts either notionally or physically. And this, judging from experience and by th definition of man, obviously is not the case.

Lesson Seven

76. In the *In VI Metaph.*, less. 2, n. 1177, this reference to Plato is elucidated St. Thomas says:

> But that accidental being is as though being in name only, Aristotle proves in two ways: first, by the authority of Plato, and secondly, through an argument.... He says, therefore, that because of the fact that accidental being is in a way being in name only, Plato, therefore, did not do badly when, in the course of orienting different sciences to different substances, he oriented sophistical science to non-being, since the arguments of the sophistics chiefly concern accidents, for hidden paralogisms come into being chiefly according to the fallacy of accident.

And in n. 1178, he adds, "For accidental being seems to verge toward non-being. And, therefore, sophistics, which is about the apparent and non-existent, is chiefly about accidental being."

Here, by way of parenthesis, one might question the reference in n. 1177 to "sophistical *science*." There *is* such a science. To the extent that one can *demonstrate* the rules for arguing sophistically, there is a *science* of arguing sophistically. The actual use of sophistical arguments that deviate from the process of true argumentation cannot, however, produce scientific conclusions.

Also in Plato's *Sophist*, conversation between Theaetetus and the Stranger makes clear the nature of the art of the Sophists with regard to their use of non-being as being. We read the following conversation in 240–241 of this dialogue of Plato:

> Str. A resemblance, then, is not really real, if, as you say, not true?
> Theaet. Nay, but it is in a certain sense.
> Str. You mean to say, not in a true sense.
> Theaet. Yes; it is in reality only an image.
> Str. Then what we call an image is in reality really unreal.
> Theaet. In what a strange complication of being and non-being we are involved!
> Str. Strange! I should think so. See how, by his reciprocation of opposites, the many-headed Sophist has compelled us, quite against our will, to admit the existence of not-being.... Again, false opinion is that form of opinion which thinks the opposite of the truth:—You would assent?
> Theaet. Certainly.
> Str. You mean to say that false opinion thinks what is not?
> Theaet. Of course.
> Str. Does false opinion think that things which are not are not, or that in a certain sense they are?
> Theaet. Things that are not must be imagined to exist in a certain sense, if any degree of falsehood is to be possible.
> Str. And does not false opinion also think that things which most certainly exist do not exist at all?... And in like manner, a false proposition will be deemed to be one which asserts the non-existence of things which are, and the existence of things

which are not...but the Sophist will deny these statements. And indeed how can any rational man assent to them, when the very expressions which we have just used were before acknowledged by us to be unutterable, unspeakable, indescribable, unthinkable? Do you see his point, Theaetetus?
Theaet. Of course he will say that we are contradicting ourselves when we hazard the assertion, that falsehood exists in opinion and in words; for in maintaining this, we are compelled over and over again to assert being of not-being, which we admitted just now to be an utter impossibility.

And so, since the Sophist is concerned with appearances, with falsehoods an with the unreal in the place of the real, it is easy for him to confuse being and no being and, in turn, to confuse us in this regard also.

77. Again we may turn to the *Sophist* of Plato to show that he was not i agreement with his "father" Parmenides, who said that "whatever is non-being nothing." After explaining that motion is not-rest, yet is, and that the not-beautifu is other than the beautiful, yet is, the Stranger continues in the following manne in 258:

Str. The same may be said of other things; seeing that the nature of the other has a real existence, the parts of this nature must equally be supposed to exist.
Theaet. Of course.
Str. Then, as would appear, the opposition of a part of the other, and of a part of being, to one another, is, if I may venture to say so, as truly essence as being itself, and implies not the opposite of being, but only what is other than being.
Theaet. Beyond question.
Str. What then shall we call it?
Theaet. Clearly, not being; and this is the very nature for which the Sophist compelled us to search.

> Str. And has not this, as you were saying, as real an existence as any other class? May I not say with confidence that not-being has an assured existence, and a nature of its own? Just as the great was found to be great and the bountiful beautiful, and the not-great not-great, and the not-beautiful not-beautiful, in the same manner not-being has been found to be and is not-being, and is to be reckoned one among the many classes of being. Do you, Theaetetus, still fill any doubt of this?
> Theaet. None whatever.
> Str. Do you observe that our scepticism has carried us beyond the range of Parmenides' prohibition?... Whereas, we have not only proved that things which are not are, but we have shown what form of being not-being is; for we have shown that the nature of the other is, and is distributed over all things in their relations to one another, and whatever part of the other is contrasted with being, this is precisely what we have ventured to call not-being.

Since the Stranger has arrived at the position that not-being is not nothing, it is possible that some might believe that he also held that the opposition of being to not-being is not a true opposition. In order to obviate this possibility, he hastens to add in 258–259:

> Let not any one say, then, that while affirming the opposition as not-being to being, we still assert the being of not-being; for as to whether there is an opposite of being, to that enquiry we have long said good-bye—it may or may not be, and may or not be capable of definition. But as touching our present account of not-being, let a man either convince us of error, or, so long as he cannot, he too must say, as we are saying, that there is a communion of classes, and that being, and difference or other, traverse all things and mutually interpenetrate, so that the other partakes of being, and by reason of this participation is, and yet is not that of which it partakes, but other, and being other than being, it is clearly a necessity that not-being should be. And again, being through partaking of the other, becomes a class other than the remaining classes, and being other than all

> of them, is not each one of them, and is not all the rest, so that undoubtedly there are thousands upon thousands of cases in which being is not, and all other things, whether regarded individually or collectively, in many respects are, and in many respects are not.

From the above, it is obvious that there is a not-being that is the other and thus, Plato, through the speech of the Stranger, is not in accord with the theory of Parmenides which declares non-being to be nothing absolutely.

Moreover, in asserting that not-being is not nothing but rather a kind of being, i.e., a "something," Plato obviously disbelieved in the unity of being in the Parmenidean sense.

Non-being in the *Sophist* is simply "the not such and such," the different. And this negative way of speaking of an existent is just as important as the positive way of speaking of an existent. As we have said above, not being an apple has just as much right to be considered as does being an apple, since it is to be something positively other than apple.

78. There seems to be no place where Plato speaks of magnitudes that are "indivisible by division." In *The Parmenides of Plato*, p. 38, A. E. Taylor says that there must be some sense in which the elements of a line are points that have no length, but, referring to Plato, he adds:

> There is no surviving authentic explanation of the sense in which he spoke of 'indivisible' lines, but he presumably meant that no repetition of the process of division of a line into lesser parts will yield a result which is not itself also a line, the 'points' which are the elements of lines are not to be obtained by repeated bisection as, of course, they could be if they were unit lengths.

We might well ask ourselves if it is really Plato that St. Thomas means; or if it is Plato, how can he rightly mean Plato? There is no evidence in the text of Aristotle to this theory of Plato, so the position of Plato cannot be evident to St. Thomas through the Aristotelian text. In fact, the dialogue, *Timaeus*, is probably the only work of Plato that was translated into Latin early enough for St. Thomas to have read it. And there is no mention of any process whereby magnitudes are rendered "indivisible by division" in the *Timaeus*.

Through Timaeus Plato does say, however, in the *Timaeus*, 53:

> In the first place, then, as is evident to all, fire and earth and water and air are bodies. And every sort of body possesses solidity, and every solid must necessarily be contained in planes; and every plane rectilinear figure is composed of triangles; and all triangles are originally of two kinds, both of which are made up of one right and two acute angles; one of them has at either end of the base the half of a divided right angle, having equal sides, while in the other the right angle is divided into unequal parts, having unequal sides. These, then, proceeding by a combination of probability with demonstration, we assume to be the original elements of fire and the other bodies;...but the principles which are prior to these God only knows, and he of men who is the friend of God.

And Timaeus continues to elucidate this in 54, when he says:

> There was an error in imagining that all the four elements might be generated by and into one another; this, I say, was an erroneous supposition, for there are generated from the triangles which we have selected four kinds—three from the one which has the sides unequal; the fourth alone is framed out of the isosceles triangle. Hence they cannot all be resolved into one another, a great number of small bodies being combined into a few large ones, or the converse. But three of them can be thus resolved and compounded, for they all spring from one, and when the greater bodies are broken up, many small bodies will

> spring up out of them and take their own proper figures; or, again, when many small bodies are dissolved into their triangles, if they become one, they will form one large mass of another kind. So much for their passage into one another.

And as far as the kinds of bodies are concerned, Timaeus says: "The first will be the simplest and smallest construction, and its element is that triangle which has its hypothenuse twice the lesser side."

The second and third elements are formed by combinations of this first triangle, but the fourth element or the fourth elementary figure is produced by the isosceles triangle. And thus, in a way, bodies are reduced to triangles as to their most elementary reduction. And, seemingly, these triangles are "indivisible by division," as understood here.

Then, too, through the Athenian in the *Laws*, Bk. X, 893, Plato also says that as far as when and how all things were first created:

> Clearly, they are created when the first principle receives increase and attains to the second dimension, and from this arrives at the one which is neighbour to this, and after reaching the third becomes perceptible to sense.

Francis Cornford in his *Plato and Parmenides*, p. 199, referring to the above citation, says:

> Discussing the passage, Miss A. T. Nicol writes: 'The arche is the indivisible line, the second stage the indivisible surface, the next the indivisible solid, and the last is the solid perceived by the senses. We see now why there is no mention of indivisible line in the *Timaeus*. The *Timaeus* is a myth of the physical world, and therefore has no need to go further back than the surface, the stage where in descending from the arche the third dimension becomes possible; for without the third dimension there is no sensation.'

Miss Nicol expresses the above quote in "Indivisible Lines," *Classical Quarterly*, Vol. XXX, p. 125. We may or may not accept this interpretation, but the fact remains that we have not found where Plato speaks of indivisible magnitudes, although we are told that he spoke of them often. In the *In I Metaph.*, less. 16, n. 258, St. Thomas says that Plato stated points to be "the principles and substance of all magnitudes." And St. Thomas also says that

> often Plato said that indivisible lines were the principles of lines and of other magnitudes and that the genus with which geometry was concerned was indivisible lines. Yet, through the fact that Plato states that all magnitudes are composed of indivisible lines, he does not evade stating the difficulty that magnitudes are composed of points and that points are the principles of magnitudes. For it is necessary that there be terminations of indivisible lines and these terminations can only be points. Thus, from the same argument by which it is stated that indivisible line is the principle of magnitude it is stated that a point is the principle of magnitudes.

St. Thomas, in referring to Aristotle's saying that Plato often referred to indivisible lines, may be restating information based on the sayings of the followers of Plato rather than on the works of Plato. Plato may have spoken of these indivisible lines or magnitudes indivisible by division and he may not have specifically incorporated this theory into his writings.

This theory is that of the cosmogony of the Pythagoreans, who proceeded from the positing of a monad as a principle of unity and an indefinite dyad as matter to numbers, and thence to geometrical solids.

79. In less. 3, n. 1 of the *In III De Caelo*, St. Thomas, commenting on Bk. III, chap. 1, n. 5, of Aristotle, says that Aristotle makes a special point to discuss the

opinion of Plato, viz., that all bodies were generated from, composed of and resolved into planes, because it was a popular opinion. St. Thomas, along with Aristotle, says that this opinion is not true, because it is contrary to the mathematics which supposes that a line is indivisible and, therefore, the plane cannot be composed of lines any more than a line can be composed of indivisible plants. Mathematicians

> assume that a line is length without width. Thus, planes, which have length, width but not depth, cannot come to be from lines. And so, a body, which has depth and width cannot come to be from planes. It is not right, however, that anyone should discard such mathematical suppositions unless he had arguments that are more probable than those mathematical assumptions. This is the reason why it seems that the foregoing opinion of Plato, which, without any compelling reason, discarded such assumptions, should be disproved.

In other words, according to mathematics, points do not make lines, lines do not make surfaces or planes, and planes do not make bodies or solids. According to St. Thomas, if what Plato says is true, viz., that "bodies are composed of planes and planes from lines and lines from points, it will be necessary that a part of a line is not a line." And St. Thomas reminds us here that it has been proved in the sixth book of the *In Physic.* that lines are neither indivisible nor composed of indivisibles.

In this lesson of the *De Caelo*, St. Thomas and Aristotle proceed in a manner quite different from their usual one. They first show that the opinion of Plato is contrary to mathematics and then they show that it is, therefore, contrary to the natural order. It seems that these two philosophers usually first show that something is true or false in the natural order and then show that it is true or false in the mathematical order. They give a reason for the procedure they use here. St. Thomas says:

> This is necessary, because whatever happens to be impossible with regard to mathematical bodies must also be impossible with regard to natural bodies. The reason for this is that entities are said to be mathematical by reason of their abstraction from natural things. But natural beings are related to mathematical entities by way of addition, for natural beings add to mathematical entities sensible nature and movement from which the latter abstract. Hence, it is evident that those things that are in mathematical entities are retained in natural ones and not vice versa. This is the reason why the incongruities that are against mathematical entities are also against natural ones, and not vice versa.

Then, for the remainder of the lesson, St. Thomas repeats the various arguments that show why natural bodies cannot be composed of planes. He states and explains a general argument and then he gives some particular arguments. Among these latter, he shows how, if the assumption of Plato is true, quantitative parts of bodies would have no weight, since points do not have weight, and, therefore, neither do lines which, according to the Platonists, were composed of points, nor do surfaces, because they are composed of lines, which have no weight.

And in the next lesson, viz., the fourth lesson of this third book, four of the particular arguments against this tenet of Plato are given. One of these arguments, the fourth one, says that if Plato's opinion is true, it would do away with magnitude:

> If a body is composed of planes, it will have the capacity of being resolved into surface, and by the same reasoning all magnitudes will be resolved into the first things, i.e., into points. And so, it would follow that there would be no body, but only points.

80. Here in n. 50, Aristotle is actually defending Plato against those who disagree with Plato. As far as Plato and non-being are concerned, it would seem that his understanding of accident as non-being, even though not absolutely nothing, is allied with his idea that non-sensible universals are *the* substances, and that all sensible appearances of things are as something somewhat illusory and deceptive. Thus, he would call accidents deceptive, but yet would say that they really were *something*.

The problem of the *Sophist* is that of "otherness" which is *a* non-being. The whole point of the *Sophist* is to show that non-being is not *being*, but that it is *being other*. Plato did not say that non-being was something in the sense that it is some one actually existing thing, but in the sense that non-being was not nothing. Not being an apple, then, is something in the same sense that being an apple is. Or if we consider the not beautiful, we see that it is not the being of being beautiful but it is the not being of being beautiful or the being of not being beautiful. It is other than the beautiful. It is not nothing; it is a kind of being, a "something." As we have quoted above from the *Sophist*, 258, not being "has an assured existence, and a nature of its own." In 257, the Stranger says:

> When we speak of not-being, we speak, I suppose, not of something opposed to being, but only different.... The negative particles...when prefixed to words, do not imply opposition, but only difference from the words, or more correctly from the things represented by the words, which follow them.

And so, it is not true to say univocally that Plato said that "non-being was something." This statement must be qualified.

81. Although no mention of Plato is made in the sixth lesson of the sixth book of the *In Physic.*, St. Thomas does prove there that "no continuum is composed of indivisibles," and in proving this, he obviously proves that there can be no individual magnitudes, i.e., no magnitudes that are individual in the sense of being completely one and admitting of no further division. After proving that the continuum is not composed of indivisibles in the case of magnitude, he then shows that the same is true of motion and also of time which is consequent upon motion.

In the first lesson of the sixth book, St. Thomas assumes the line as an example of a continuum and a point as an example of the indivisible. First, he proves that the continuum is not composed of indivisibles either through the mode of *continuation* or through the mode of *contact*. By definition a continuum is a quantity that has two and only two ultimate terminations. Each termination is an indivisible point.

In the first argument proving that a continuum cannot be composed of indivisibles, St. Thomas says that a continuum cannot be composed of indivisibles in the mode of *continuation*, because then the continuum would not have just one ultimate terminal point at each end of the continuum. There would be many points. In a continuum, each of the two terminal points is not just a one in itself, since it indicates a relation to the whole. It is a termination of something. The last points, the terminal points, moreover, are not co-existent with the whole, for the point is one thing and that of which it is a point is another. If indivisibles were to compose the continuum, there would be no ultimate point in indivisibles that could be a part, because it is impossible that there be a part in an indivisible. It is of the nature of indivisibles that they have no parts. And so, a continuum, e.g., a line, cannot be composed of indivisibles by way of or in the mode of *continuum*.

Nor can a continuum be composed of indivisibles in the mode of *contact* or touch, because, if this were the case, the two ultimate points of two indivisibles would have to be together. This is impossible, because in indivisibles, as we have said, there are no parts. And the ultimate of an indivisible would have to be the ultimate of something. And to speak of an ultimate and that of which it is an ultimate is to speak of parts, which in this case is impossible, because it is against the very nature of an indivisible to have parts.

St. Thomas states the second argument in n. 753:

> If a continuum is made up of points, either they must be continuous with one another or they must touch one another. And the same argument holds for all other indivisibles, because a continuum is not composed of them.
>
> But the first argument is sufficient to prove that indivisibles cannot be mutually continuous.
>
> But to prove that they cannot touch each other, he introduces another argument. It is as follows: In the case of everything which touches another, either one whole touches another whole, or a part of one touches a part of the other, or a part of one touches the other whole. But, since the indivisible does not have a part, it cannot be said that a part of one touches the part of another, or that a part touches the whole; and so, it is necessary, if two points touch each other, that the whole touch the whole. But a continuum cannot be composed of the two, of which one whole touches another whole, because every continuum has parts that can be separated in such a way that this is one part and that another; and it is divided into different and distinct parts in *place*, i.e., in position in these which have position. But those which touch according to the whole are not distinguished in place or position. It remains, therefore, that a line cannot be composed of points in the mode of touch (contactus).

He goes on to give another reason why a continuum cannot be composed of indivisibles, but this second reason that we have given above seems more cogent.

We have to add again that, although St. Thomas must be considering the content of the first lesson of the sixth book of the *In Physic.* here in the seventh lesson of the first book of that work, the former does not seem pertinent for any refutation of Plato, for we do not know the locus in which Plato has stated his belief in the indivisibility of magnitudes.

82. As we have said elsewhere, being, for St. Thomas, was not a genus. And Plato, in the *Sophist*, 254–255, states through the Stranger that there are important *classes* which may or may not have communion with each other, but that all these classes have communion with being:

> The most important of all the genera are those which we were just now mentioning—being and rest and motion.... And two of these are, as we affirm, incapable of communion with one another.... Whereas being surely has communion with both of them, for both of them are?

And, as usual, Theaetetus answers, "Of course." They go on to discuss the other two classes, viz., same and other, but show that being somehow is the foundation of all. They do not, however, make lucid the nature of the doctrine of "communion," although that there is such a doctrine is certainly set forth. Richard Robinson, *op. cit.*, p. 263, referring to this theory, says that Plato "does not tell us what it is, but only *that* it is."

When asked what he means by the "nature of being," the being in which all classes communicate, the Stranger had already said in the *Sophist*, 247–248.

My notion would be, that anything which possesses any sort of power to affect another, or to be affected by another, if only for a single moment, however trifling the cause and however slight the effect, has real existence; and I hold that the definition of being is simply power.... Let us now go to the friends of ideas; of their opinions, too, you shall be the interpreter.

To them we say—You would distinguish essence from generation?

Theaet. 'Yes,' they reply.

Str. And you would allow that we participate in generation with the body, and through perception, but we participate with the soul through thought in true essence; and essence you would affirm to be always the same and immutable, whereas generation or becoming varies?

Theaet. Yes; that is what we should affirm.

Str. Well, fair sirs, we say to them, what is this participation, which you assert of both? Do you agree with our recent definition?

Theaet. What definition?

Str. We said that being was an active or passive energy, arising out of a certain power which proceeds from elements meeting with one another.... Any power of doing or suffering in a degree however slight was held by us to be a sufficient definition of being.

Theaet. True.

Str. They deny this, and say that the power of doing or suffering is confined to becoming, and that neither power is applicable to being.

Theaet. And is there not some truth in what they say?

Str. Yes...but they will allow that if to know is active, then, of course, to be known is passive. And on this view being, in so far as it is known, is acted upon by knowledge, and is therefore in motion; for that which is in a state of rest cannot be acted upon, as we affirm....

And, O heavens, can we ever be made to believe what motion and life and soul and mind are not present with perfect

> being? Can we imagine that being is devoid of life and mind, and exists in awful unmeaningness an everlasting fixture?
> Theaet. That would be a dreadful thing to admit, Stranger.
> Str. But shall we say that being has mind and not life?... Or shall we say that both inhere in perfect being, but that it has no soul which contains them?... Or that being has mind and life and soul, but although endowed with soul remains absolutely unmoved?
> Theaet. All three suppositions appear to me to be irrational.
> Str. Under being, then, we must include motion, and that which is moved.

It seems quite clear from the above that all things that are participate in the substrate, being. According to both Plato and St. Thomas, nothing is unless it shares in being. There is no motion and no rest unless there is being in which motion and rest participate.

83. Nor does Plato say that accident or not being is non-being in the full sense. He makes "not-being" to be "the other," and his whole point was to show that his "not-being" was not nothing or non-being strictly speaking. We fail to see the problem that has been created here by the imposition of the Aristotelian or Thomistic frame of reference upon a Platonic tenet that is not Platonic.

84. This refers to n. 44 in the previous lesson, where St. Thomas says that Aristotle shows that Parmenides does not conclude properly that being is absolutely one. The reason why he does not so conclude is the multitude of parts, both the *quantitative* parts and the parts of the *definition*. In nn. 44 ff., he shows that if this "one" is not *quantitatively* divisible, it will not have any magnitude and, therefore, will not be able to be in a corporeal substance. Then he shows that it would not be *definable*, since the parts of a definition of a substance, e.g., "animal" and

"biped" of "man," are in the *substantial* order (they cannot be *accidents*) and ar many. The "one" would, therefore, have to be absolutely indivisible, i.e., it coul have neither quantitative parts and magnitude nor parts of a definition. He said th at the end of n. 45, where he said: "So, therefore, it is evident that, if only or being is stated, neither quantitative parts nor parts of magnitude nor parts of th definition can be stated."

In this passage quoted above, Aristotle and St. Thomas are concerned wi showing that Parmenides "did not conclude properly," in the sense that h conclusions would involve denying quantitative parts and magnitude, as well a parts of a definition. And this is contrary to what appears. (We must keep mind here that Aristotle is not attacking Parmenides and Melissus *ex profess* This is the function of first philosophy, but here Aristotle is simply bringing o their incongruities.)

Here in n. 52, St. Thomas and Aristotle are simply stating that *Plato ga in too easily*. Plato said that if non-being, which is an accident, is not made to b something, then, all things are one. But St. Thomas and Aristotle say that even it is supposed that there is only substance, and that accident, indeed, was nothin it still would not follow that being was one. In effect, even forgetting the divisi of substance into quantitative parts, it would still have to have defining parts, su as "animal" and "biped" in the case of "man." And these parts are in t substantial order.

Furthermore, substance, while being *one* generically, would still be *multip* in species, and this Plato would have to hold, since he maintained that there we many species. In *In VII Metaph.*, less. 14, n. 1592, St. Thomas begins to spe of the incongruities resulting from Plato's saying that "ideas are substances a separable and are said to be universal species" and from his saying along with this that "species were made up of genus and differences."

St. Thomas goes on to say that besides this division of the substance man into other things that are also of the genus substance, there are also many actual substances that differ on the basis of the various differences of the genus. And again things are not just one but many.

Lesson Eight

85. Plato has referred to the great and the small in two ways. In the early dialogues, he considers them as norms by which the great is great and the small is small. This is discussed in the *Statesman*, 283–285. The Stranger says to Young Socrates, "Do you not think that it is only natural for the greater to be called greater with reference to the less alone, and the less less with reference to the greater alone?"

After Young Socrates has answered affirmatively, the Stranger continues:

> Well, but is there not also something exceeding and exceeded by the principle of the mean, both in speech and action, and is not this a reality, and the chief mark of difference between good and bad men?

And again after Young Socrates gives another affirmative answer, the Stranger says:

> Then, we must suppose that the great and small exist and are discerned in both these ways, and not, as we were saying before, only relatively to one another, but there must also be another comparison of them with the mean or ideal standard; would you like to hear the reason why?

And after still another affirmative answer from his listener, the Stranger continues in this bit of dialectical obstetrics through which he hopes to make clear the nature of great and small as relative to a *mean*.

> If we assume the greater to exist only in relation to the less, there will never be any comparison of either with the mean.... And would not this doctrine be the ruin of all the arts and their creations; would not the art of the Statesman and the aforesaid art of weaving disappear? For all these arts are on the watch against excess and defect, not as unrealities, but as real evils, which occasion a difficulty in action; and the excellence of beauty of every work of art is due to this observance of measure.
>
> ...Well, then, as in the case of the Sophist we extorted the inference that not-being had an existence, because here was the point at which the argument eluded our grasp, so in this we must endeavor to show that the greater and the less are not only to be measured with one another, but also have to do with the production of the mean; for if this is not admitted, neither a statesman nor any other man of action can be an undisputed master of his science.
>
> ...I think...that we may fairly assume nothing of this sort.
>
> ...That we shall some day require this notion of a mean with a view to the demonstration of absolute truth; meanwhile, the argument that the very existence of the arts must be held to depend on the possibility of measuring more or less, not only with one another, but also with a view to the attainment of the mean, seems to afford a grand support and satisfactory proof of the doctrine which we are maintaining; for if there are arts, there is a standard of measure, and if there is a standard of measure, there are arts; but if either is wanting, there is neither.
>
> ...The next step clearly is to divide the art of measurement into two parts, as we have said already, and to place in the one part all the parts which measure number, length, depth, breadth, swiftness with their opposites; and to have another part in which they are measured with the mean, and the fit, and the opportune,

> and the due, and with all those words, in short, which denote a mean or standard removed from the extremes.

Now the Young Socrates seems to sum up the whole point, when he says: "Here are two vast divisions, embracing two very different spheres." And as if in confirmation of this statement, the Stranger replies:

> There are many accomplished men, Socrates, who say, believing themselves to speak wisely, that the art of measurement is universal, and has to do with all things. And this means what we are now saying; for all things which come within the province of art do certainly in some sense partake of measure. But these persons, because they are not accustomed to distinguish classes according to real forms, jumble together two widely different things, relation to one another, and to a standard, under the idea that they are the same, and also fall into the converse error of dividing other things not according to their real parts. Whereas the right way is, if a man has first seen the unity of things, to go with the enquiry and not desist until he has found all the differences contained in it which form distinct classes; nor again should he be able to rest contented with the manifold diversities which are seen in a multitude of things until he has comprehended all of them that have any affinity within the bounds of one similarity and embraced them within the reality of a single kind. But we have said enough on this head, and also of excess and defect; we have only to bear in mind that two divisions of the art of measurement have been discovered which are concerned with them, and not forget what they are.

In the *Republic*, Bk. VII, 524, it is stated that the soul, in knowing, uses other means than the senses. When considering whether sight adequately perceives smallness and greatness of fingers, Socrates says:

> The eye certainly did see both small and great, but only in a confused manner; they were not distinguished.... Whereas the thinking mind, intending to light up the chaos, was

> compelled to reverse the process, and look at small and great as separate and not confused.... Was not this the beginning of the enquiry 'What is great?' and 'What is small?'... And thus arose the distinction of the visible and the intelligible.

Again in the *Phaedo* we find the same theme. In 96, Socrates tells Cebes that he used to think that he understood what greater and less meant. He says that when he saw a tall person standing next to a short person, he thought that it was safe to assume that the tall one was greater than the short one. Then Socrates goes on to explain that after careful study he realized that one man was not taller than another "by a head" but by participation in greatness. After agreeing "that by greatness only great things become great and greater greater, and by smallness the less become less," Socrates says in 101:

> Then if a person were to remark that A is taller by a head than B, and B less by a head than A, you would refuse to admit his statement and would stoutly contend that what you mean is only that the greater is greater by, and by reason of, greatness, and the less is less only by and by reason of, smallness; and thus you would avoid the danger of saying that the greater is greater and the less less by the measure of the head, which is the same in both, and would also avoid the monstrous absurdity of supposing that the greater man is greater by reason of the head, which is small.

In line with the procedure by which the *Parmenides*, *Sophist*, *Statesman* and *Philebus* criticizes the position of the *Republic* and the *Phaedo*, the theory of the great and small as relative to a norm is criticized in the *Parmenides*. In 131, Parmenides asks Socrates if, because of his youth and inexperience in philosophy, he believes that "great things become great because they partake of greatness." And Socrates answers affirmatively.

Thus far, what we have said of this meaning of great and small as understood y Plato is that they are relative to a norm or mean which is the measure by which omething is great or small or by which something has a kind of excess or defect.

Plato considered the great and small in a *second way* and this is the way that ; most pertinent here, for it concerns the great and small precisely as principles of eing. Actually, however, it is only in the middle dialogues, and to a certain xtent in the *Timaeus*, that the point that is made here with regard to the great and mall as principles really holds, for it would seem that these are the only dialogues ıat St. Thomas ever knew. Moreover, we find no indication in the mature Plato vhich says that the great and small are principles.

Considered as such, i.e., as principles of being, the great and small are often alled the "indeterminate dyad." It is in this second consideration that the great nd small are as the unlimited. They are somewhat like Aristotelian prime matter, vhich is unlimited and indefinite until it is limited or made definite by form. It has o dimension. Plato's great and small are also indeterminate. They are indeter-ıinately great and small. They are, in a sense, "not anything definite or in articular," and, therefore, are comparable to what prime matter was in the ıeorizing of Aristotle.

This duality of great and small which Plato stated to be on the side of matter ; also mentioned by St. Thomas in the *In VII Physic.*, less. 6, n. 938:

> That the diversity of things depends upon the diversity of the receptive only is the Platonic opinion, for this opinion stated unity from the viewpoint of form, and duality from the viewpoint of matter in such a way that the whole notion of diversity comes from the material principle. Hence, he stated that *one* and *being* are said univocally and signify one nature, but that the species of things are diversified according to the diversity of what are receptive.

And in less. 8, n. 955, St. Thomas is even more specific in referring to this "indeterminate dyad," the great and small, which are considered to be principles of being according to Plato:

> And thus it is that Plato stated *one* as a species, but he stated *great* and *small*, which are on the side of matter, as the contraries through which things are diversified. And thus, it will follow that just as one and the same health has a duality to the extent that it receives the more and the less, so too, substance, which is number, since it belongs to one species on the side of unity, will have a duality to the extent that the number is greater or less. But in the case of substances, there is stated no common term which will signify both, i.e., the diversity which occurs from the greater and the less of number.

This indefinite dyad, the great and the small, the indefinite something, is pointed up in 158 of the *Parmenides*, where, speaking of what does not participate in the idea of the One, Parmenides, seemingly the mouthpiece of Plato, says:

> And if we continue to look at the other side of their nature, regarded simply, and in itself, will not they, as far as we see them, be unlimited in number?... And yet, when each several part becomes a part, then the parts have a limit in relation to the whole and to each other, and the whole in relation to the parts.... The result to the others than the one is that the union of themselves and the one appears to create a new element in them which gives to them limitation in relation to one another; whereas in their own nature they have no limit.... Then the others than the one, both as whole and parts, are infinite, and also partake of limit.... Then they are both like and unlike one another and themselves.... Inasmuch as they are unlimited in their own nature, they are all affected in the same way.... And inasmuch as they all partake of limit, they are all affected in the same way.... But inasmuch as their state is both limited and unlimited, they are affected in opposite ways....

If the element of unity is removed from things, these things are then both 'eat and small, unlimited, indefinite. A part of this indefinite something can by articipation in unity become a definite something. Is this not similar to the theory prime matter, which, by being informed by some form, can become something particular, i.e, a definite something from an indefinite something? It seems quite isy to see why Aristotle said that the great and small, i.e., the indefinite dyad, as the material element in the becoming of things. It was the element of iversity, while the form was the element of unity. Contrariety was found in the aterial and not in the formal element. The material element, the indefinite dyad, what is not one, i.e., what is other than one, but which, by partaking in One, an become something definite.

Again in 24 of the *Philebus* Plato through Socrates speaks of the classes of ie definite and the indefinite or the finite and the infinite as principles of the ecoming of things. Socrates says to Protarchus:

> The two classes are...the finite, and...the infinite; I will first show that the infinite is in a certain sense many, and the finite may be hereafter discussed.... And now consider well; for the question to which I invite your attention is difficult and controverted. When you speak of hotter and colder, can you conceive any limit in those qualities? Does not the more and less, which dwells in their very nature, prevent their having any end? For if they had an end, the more and less would themselves have an end.

And after Protarchus affirms the above, Socrates continues, "Ever, as we ay, into the hotter and the colder there enters a more and a less.... Then, says the rgument, there is never any end of them, and being endless they must also be nfinite."

Protarchus says that this is "exceedingly true." And then Socrates ente into a very clear discussion of the definite and the indefinite:

> Yes, my dear Protarchus, and your answer reminds me that such an expression as 'exceedingly,' which you have just uttered and also the term 'gently,' have the same significance as more or less; for whenever they occur they do not allow of the existence of quantity—they are always introducing degrees into actions, instituting a comparison of a more or a less excessive or a more or a less gentle, and at each creation of more or less, quantity disappears. For, as I was just now saying, if quantity and measure did not disappear, but were allowed to intrude in the sphere of more and less and the other comparatives, these last would be driven out of their own domain. When definite quantity is once admitted, there can be no longer a 'hotter' or a 'colder' (for these are always progressing, and are never in one stay); but definite quantity is at rest, and has ceased to progress. Which proves that comparatives, such as the hotter and the colder, are to be ranked in the class of the infinite.

Then Socrates very concisely states that what admits of a more or a less i placed together under the division or class called the "infinite" and that what doe not admit of a more or a less, but which admit their opposites, e.g., equal, i placed under the division or class called the finite or limited.

In this last citation, viz., that from the *Philebus*, the similarity of Plato' indefinite plasticity to Aristotle's prime matter and of Plato's limiting idea t Aristotle's form seems clearer than in the other citations. It should now be quit apparent that the "great and small," the indefinite, unlimited element is taken a from the viewpoint of matter, while the unifying element is the formal element The unlimited range of the great and small is limited by some formal element o unity. Until this limitation takes place, the material element is composed o contraries, e.g., great-and-small, that are not yet made definite. The materia ement is both great and small, or greater and smaller, if you will. The multi-icity from the viewpoint of the matter gains unity from the viewpoint of the form.

Cf. also Appendix note 126.

esson Nine

5. Here as elsewhere, when St. Thomas and Aristotle refer to the natural ilosophers it is well to know that they call them natural philosophers not only ecause they tried to explain the essence of things by a natural principle but also ecause they tried to explain all the species of movement in nature.

7. For Anaxagoras it seemed that contraries always came from each other, as ack from white and vice versa, and so, he said that contraries came from each her. Although we see that something cold may become warm, we cannot say that e essence coldness becomes the essence heat. Where there is movement, it is om one contrary to the other in the same subject.

8. Since Anaxagoras says that there must be an infinite number of anything hatsoever in anything whatsoever, there must be an infinite number of infini-simal parts of flesh in non-flesh.

9. In other words, it is the predominance of minimal parts of flesh or of wood at determines the nature of a given thing.

0. We cannot know infinity as divisible into an actual infinity of parts, because finity cannot be so divided. Knowledge of infinity in this way is inaccessible to

us not only because it is intellectually impossible for us to comprehend this, b
also because such a division is physically impossible. Division into an actu
infinite number of parts would destroy the nature of infinity, for by definition,
"is that of which there is always something beyond," as is said in the *In P*
Physic., less. 11, n. 383. It is just as impossible for us to know the infini
divisions of anything as it is for us to know future contingents as future. Th
following division may prove helpful for the understanding of the various kinds o
infinity:

1. *Material*: numerical infinity within a species. This means that there may be an infinite number of members within a given species. Here, the matter may be either the continuum or a series of numbers.
2. *Formal*: infinity according to the form in a limited being. The form of an angel is substantially infinite, i.e., non-finite in its own order, because it is not participated in or limited by reception in matter. The same subject is also substantially finite in the order of *existence*, because its being (esse) is received by the form.
3. *Infinity as such* or in the full sense. Being (esse) exists in subsistent plentitude unparticipated or received by essence or form. There is no limitation in any order.

In the *Summa Theol.*, Ia, qu. 7, art. 1, we find that the relationship betwee
form, matter and infinity is explained in a clear manner. St. Thomas says:

> We must say that 'all the ancient philosophers attribute infinite to the first principle,' as is said in *Physics* III, and 'this with reason,' considering that things flow to infinity from a first principle. But because some erred about the nature of the first principle, they consequently erred about its infinity, for, since they stated that the first principle was matter, they consequently attributed to the first principle material infinity when they said that some infinite body was the first principle of things.

> We must consider, therefore, that something is called infinite from the fact that it is not finite. But in some sense both the matter is limited by the form and the form by the matter. The matter, indeed, is limited by the form to the extent that the matter, before it receives the form, is in potency to many forms, but when it receives one, it is terminated by it. But the form is limited by the matter to the extent that the form, considered as such, is common to many. But by the fact that it is received in matter, it is determinately made the form of this thing. The matter, however, is perfected by the form that limits it. And, therefore, the infinite, as attributed to matter, has the characteristic of imperfection, for it is as matter not having form. But form is not perfected by the matter but rather its amplitude is contracted by it. And so the infinite, considered from the viewpoint of form not determined by matter, has the characteristic of perfection.

And he adds:

> But what is the most formal of all is being itself.... Since divine being is not received in anything but is itself its own subsistent being...it is, therefore, clear that God Himself is infinite and perfect.

Briefly, then, we may say that *matter* is:

1. Determined by the form.
2. Limited by the form.
3. Perfected by the form.
4. That of which imperfect infinity can be said.

And we may say of *form* that it is:

1. Not determined by the matter.
2. Limited by matter because it becomes the form of the thing.
3. Not perfected by the matter but derives its limitation from being in matter.

4. The subject of perfect infinity. Considered in itself, it is infinite in the sense that it has no limitation of itself. It is itself completely what it is actually.

91. It is obvious, if the nature of the principles that are infinite are unknown tha the things composed from such principles will also be unknown, for we know a composite only when we know the nature of the parts that compose it. In less. 1 n. 5 of the *In I Physic.*, St. Thomas says that we have knowledge in the order o natural philosophy when we know the *principles* and *causes* and *elements*. In the case of principles of natural things as they are considered here as the element o Anaxagoras, these principles are the elements *and* a cause of a thing's being Obviously, then, if the principles are unknown by reason of their being infinite and therefore, incomprehensible to the human intellect, what is composed of these unknown principles or elements will also be unknown.

92. This is another instance of the question of whether or not we know a thing before we know the complete composition of the thing.

93. The parts of a line do not actually exist prior to the existence of the line, bu they only potentially exist as parts after the line exists. A line is a line in act, bu it is only potentially divisible into parts.

In the *In VII Physic.*, less. 9, n. 960, St. Thomas says:

> If any part, whatever be its size (quantacumque), that exists in a whole moves, the part does not have to be able to move separately as existing in the full sense (per se existens). The reason for this is that a part which is in a whole is not in act, but in potency, especially in the case of continua. For something is being in the way that it is also *one* (aliquid est ens,

sicut et unum); but *one* is what is undivided in itself and divided from all others. A part, however, according as it is in a whole, is not divided in act, but in potency only. Hence, it is neither being in act nor is it one in act, but only potentially. And for this reason, also, the part does not act, but the whole does.

4. If the whole has a certain size and is composed of actual parts, the size of the vhole cannot be indifferent. If one builds a house with so many bricks of such a ize, the size of the house may not be indifferent, i.e., the house cannot be nfinitely small or infinitely great.

The necessity of a given matter is derived from the purpose which the matter s to serve. Matter has a certain nature and this nature implies certain conse-uences. To build a house, one cannot use any matter whatsoever, but only a ertain kind of matter. One cannot build a house of feathers only, although one night make a house by using feathers in reinforced cement. We are, of course, peaking of a house for human habitation.

The finality of matter is ordered to the form. What is true of matter is true f quantity, for in the last analysis, the finality of quality is the form. A given orm is the form of a given quantity, and not of quantity indifferently, for to be a iven matter it must be an already informed matter. That there should be such xigencies is due to the nature of matter and to the nature of quantity.

The total being of matter is ordination to form. There is an ordering of natter to form; there is successive information. We must note *very carefully* that vhen we speak of prime matter's relation to form, this is merely in the order of nalyzing, for prime matter in this sense is the fundamental material component of n already existing being. It is not the indefinite capacity to form of prime matter n general. The being of matter is being-for-form. But a form that informs matter does not do so indifferently in the sense of informing any matter whatsoever. I the order of change, matter is a relationship to all forms but not to all form directly. It is not in potency to all forms but is prepared for a form by the forr which it has.

Both a natural and an artificial form require a matter appropriate to tha form. The form "treeness" could not inform the matter that the form "grass hopper" informs. Nor can the artificial form "house" be imposed upon the matte of feathers. The latter form may be imposed upon the matter of wood and brick because of the nature of these kinds of matter. But a house of certain specification needs a certain and not an indefinite number of bricks, since this is material o sensible quantity. As such, it has "parts outside of parts" and occupies a definit amount of space. Moreover, according to St. Thomas, the rule of making i artificial things states that the end or work to be accomplished is open to determina tion by the artist but that once this goal or end has been established, determine means must be employed. And these means or media, in the case of sensibl matter, have quantity, which, by its very nature demands the consideration an realization of spatial specificity.

In short, bricks are made in view of the house, but to make the house on needs a definite quantity of bricks, and also a definite number of bricks. Th number depends upon the size of the bricks and the size of the house that on intends to build. We cannot make anything whatsoever from anything whatsoever nor can we make an entity of indefinite size.

95. It seems that perceptibility is made a condition of a thing's existence in that if something is too small to be seen, then it does not exist. The infinitesimal and imperceptible, however, do not base their being what they are upon our perceiving

them. Things have a being of their own, independent of our knowledge of them. Awareness on the part of a knowing subject cannot cause a certain matter and a certain form to exist as a certain individual being. Not only would this annihilate objectivity and the objective foundation of knowledge in the Thomistic epistemological tradition, but it would invalidate the whole study of the philosophy of nature from the Aristotelian-Thomistic point of view. Our knowledge of a thing is not responsible for a thing's act of being or for its essence. Our perception of things is not a condition of their being. To believe this would be to go against the very nature of things, as Berkeley did.

96. This is apparent from the argument above only *if* we assume the proposition of Anaxagoras to be a true one. But equal parts infinite in number in a finite being cannot be divisible to infinity. And so what Anaxagoras said, viz., that every finite body is extracted or cut off from a finite body, is not true. And if, contrary to what Anaxagoras says, some minimal part of flesh is necessary, since when something is removed, every body becomes less because the whole is greater than the part, then, nothing can be segregated from this minimal part. And thus, the tenet of Anaxagoras which says that anything at all comes to be from anything at all is false. If it were true, it would follow that infinite bodies were infinitely in infinite bodies, which is false, because it is impossible.

Besides, if we proceed logically and admit with Anaxagoras that there is no smallest part but always a smaller, there is a correlation in the divisibility to infinity of a part howsoever small and of a part howsoever large, for the same line of reasoning holds for the large being larger. A tiny line, the smallest possible, would be larger than its parts and smaller than another part. The argument for the larger parts would be the same as that for the smaller ones.

97. The reason for this is that if segregation stops, there will be a minimal part from which nothing more can be segregated and so anything will no longer be in anything. Or, if segregation stops because there is no longer anything to segregate from, obviously, there will be nothing and anything will not be in anything, because there will be no "anything" in which it may be.

98. This impossibility of having finite parts infinitely removed from a finite body is based upon the assumption that the parts removed are equal.

99. This is another way of enunciating the old Aristotelian principle that a potency is not disposed in any way whatsoever, but in the way that it is meant to be disposed. Not any thing whatsoever comes from any other thing whatsoever. Something can only come from that which is in potency to it.

100. In the Wicksteed and Cornford translation of the *In Physic.*, p. 48, Cornford tells us to take the meaning of separated particles as "the bits of *any* one substance ...separate from other bits of the *same*" substance.

101. In the *In I Metaph.*, less. 4, n. 90, St. Thomas tells us that, as far as the manner of Anaxagoras' statement is concerned, Empedocles' manner of speaking was superior to that of Anaxagoras. This is because Empedocles posited fewer principles than Anaxagoras. Speaking of Anaxagoras, St. Thomas says that "he said that material principles were infinite." And St. Thomas continues:

> It is nobler to take finite and fewer principles, as Empedocles did, as is said in the first book of the *Physics*. For, not only did he say that the principles of things were fire, water

> and other elements, as Empedocles did, but he stated as principles of things the infinite minimal parts of all things that have co-similar parts, as flesh, bone, marrow and like things. And he stated that in any one thing there were finite parts of singulars because of the fact that in inferior things we find that one thing can be generated from another, although he said that there was generation of things only through separation from the mixture, as he explained more clearly in the first book of the *Physics*.

And in n. 91 we read:

> Anaxagoras also agreed with Empedocles in that the generation and corruption of things was accomplished only through the coming together and segregation of the foregoing infinite parts and that otherwise nothing was generated or corrupted. But he said that the infinite principles of things of this kind from which substances were produced remained sempiternal.

Anaxagoras is less to be praised than Empedocles from the viewpoint of the reat number of principles that he posited. Yet, there are two noteworthy points ıat Anaxagoras makes. One is that generation came to be only through separation ·om the mixture, and the other is the fact that substances were sempiternal. These vo points were not common to all the opinions of the natural philosophers.

02. There are, however, more specific texts in the *In Metaph.* which tell us in hat sense the opinion of Anaxagoras was true and in what sense not. These texts re in less. 12, nn. 194–199, of the first book. The criticism of Anaxagoras is ery important for our purpose because it was another attempt to discover a closer pproach to the true principles of natural beings. These principles had been sought ɔr centuries and are studied in detail in this book. For these reasons we shall

present these important texts in translation. The reader will gain from them a appreciation not only of the difficulty of the task of discovering these principl but also of the detailed probing of the ancients to discover them. And, as we hav implied, there is also a good criticism and analysis of the methodology and theor of Anaxagoras as presented by Aristotle.

At the beginning of n. 194, St. Thomas says that Aristotle first showed ho the opinion of Anaxagoras was true and how it was false *in general*:

> First he says, therefore, that if anyone wanted to accept the opinion of Anaxagoras stating that there are two principles, viz., matter and an agent cause, he should also accept the reasoning that pushed Anaxagoras to express such an opinion, for Anaxagoras seems to have been forced by the necessity of truth, preoccupied as he was to follow those who had given that reason. But he himself 'never stopped to articulate that reason.' His opinion, therefore, is true in that it presupposes but does not say, but false in what it says.

And in n. 195, he shows how the opinion of Anaxagoras is true or fals *specifically*.

> For, if his opinion were considered as a whole, according to what was superficially apparent from what he said, a greater absurdity will appear for four reasons.
>
> 1. First, this very fact, viz., that all things were mixed in the beginning of the world, is absurd, since, in the opinion of Aristotle, the distinction of the parts of the world is considered to be sempiternal.
> 2. The second reason is that the unmixed is related to the mixed as the simple to the composite: but simple things exist before the composites, and not conversely: therefore, the unmixed must exist before the mixed. Anaxagoras stated the contrary of this.
> 3. The third reason is that anything at all is not by nature to be mixed with anything at all in bodies. But only those things

are by nature to be mixed which by nature are to change into one another through some alteration so that the mixture may be a union of changed things that can be mixed. But Anaxagoras stated that anything at all is mixed with anything at all.

4. The fourth reason is that there is mixture and separation of the same things, for only those things are said to be mixed which are by nature fit to exist as separated. But passions and accidents are mixed with substances, as Anaxagoras said. It would follow, therefore, that passions and accidents can be separated from substances. This is obviously false. Therefore, those absurdities appear, if the opinion of Anaxagoras is considered on the surface.

And he continues in n. 196 where he says:

Nevertheless, if anyone would 'articulately' pursue, i.e., if he would distinctly and clearly investigate what Anaxagoras 'wants to say,' i.e., what his intellect strove for, even though he did not know how to express it, what he said will appear to be more wonderful and more subtle than what the preceding philosophers said.

And this is so for two reasons: *first*, because he came closer to the true cognition of matter. And this is clear from the fact that in that mixture of things when there was nothing apart from another but all things were mixed together, nothing true could be predicated of that substance mixed in this way, which he stated to be the matter of things, as is clear in the case of colors. For, no special color could be predicated of it so that it would be said to be white or black or colored according to any other color, because according to this that color *would* have to be mixed with the others. And similarly color generically could not be predicated of it so that it might be said to be colored, because for the genus to be predicated of some one thing some species of it must be predicated, whether it be a univocal or a denominative predication. And so, if that substance were colored, it would of necessity have some determinate color, which is contrary to what was said above. And the reasoning

> is similar for 'humors,' i.e., tastes, and for all other things of this kind. Nor could first genera themselves be predicated of it, viz., to be of such a kind or of such a size or something of this sort. For, if genera were predicated, some one of the species of particular thing would have to be in it. And this is impossible, if all things are stated to be mixed together, because that species, which would be said of that substance, would already be distinct from the others. And this is the true nature of matter, viz., that it does not have any form in act but that it is in potency to all forms, because the thing itself that is mixed also does not have in act any of those things that belong to its mixture, but only potentially. He seems to have posited the foregoing mixture on account of this similarity of prime matter to the mixed thing, although there is a difference between the potency of matter and the potency of the mixed thing. For, things that can be mixed, although they are in potency in the mixed thing, however, are not in it in potency that is purely passive, for they remain virtually in the mixed thing. And this is clear from the fact that the mixed thing has motion and operations by virtue of the bodies that can be mixed. And this cannot be said of these things which are in potency in prime matter. There is also another difference, because the mixed thing, although it is not in act any one of the things that are mixed in it, nevertheless, is something in act. This cannot be said of prime matter. But Anaxagoras seems to remove this difference by the fact that he did not posit any particular mixture but a universal mixture of all things.

And in n. 197, he continues:

> In the second place, he spoke more subtly than the others because he came closer to the true cognition of the first agent principle for he said that all things were mixed by an intellect and he said that this alone was unmixed and pure.

In the two brief numbers that follow we find a good summary of the position of Anaxagoras with regard to the intellect and prime matter. In n. 198 we read:

> From these it is clear that he stated that there were two principles. And he stated that the intellect itself was one and it is accordingly simple and unmixed. And the other principle that he stated was prime matter which we state as indeterminate before it is determined and before it shares in some species, for matter, since it is of infinite forms, is determined through form, and through it it achieves some species.

And in n. 199 St. Thomas says:

> It is clear, therefore, that Anaxagoras according to what he expressed spoke neither correctly nor fully. Nevertheless, he seemed to say directly something closer to his successors' opinions which are more true, viz., to the opinions of Plato and of Aristotle who thought correctly about prime matter. These latter opinions were, indeed, more apparent at that time.

There seems little that one need add after this lengthy interpretation by St. homas of what Aristotle said of the work, aims, accomplishments and failures of naxagoras. We should say, however, that Anaxagoras seems thus far to be a reat precursor of the Aristotelian-Thomistic theory of hylomorphism. (This theory ill be expounded in the following chapter.) In addition to praising him for his roximity to the theories of prime matter and of an intellectual agent, he should so be praised for his treatment of infinity and the limit. His method of the limit nd of infinity are found closely bound to one another and he does not posit one to ie exclusion of the other.

)3. It is of the very nature of necessary accidents that they are "per se" united their subjects and are, therefore, inseparable from them in the order of kistence. Moreover, as St. Thomas says, if we push the doctrine of segregation its ultimate issue, we would have necessary accidents without subjects in which they might inhere. And also, if whiteness is removed from a subject or from mixture, something else takes its place, and so really there is no removal in th strict sense, although there may be the removal of whiteness. If sweet is remove something else takes its place and so on to infinity. We call attention again to th fact that here we are speaking of accidents that are necessary to a subject and n of separable accidents. And if we keep this in mind, we realize that segregation not possible when a mixture includes this kind of accident.

104. What is noteworthy here is that in this conception Anaxagoras saw that it w proper to the intellect to order. Further, in posing this intellect as a first principl he posed a really first principle, unmixed, uncomposed, and simple.

105. The intellect would certainly be "unbecoming" or "unfit" if it tried achieve total segregation in things that are mixed together. It would be doing wh is not its function.

Quantitatively, the continuous is divisible to infinity, but we cannot attain th *point* by total division. It is impossible to segregate all qualitative modification i.e., accidents, because some accidents are by their very nature inseparable fro their subjects. Besides, if we take away the color white from an object, some oth color would be present and so on to infinity.

Although the intellect as posited by Anaxagoras is a separated orderin intelligence and is better than the first efficient principle or cause as posited by h predecessors, Plato seems to consider it as a "deus ex machina." This, indee although no supreme compliment, is no reproach. It certainly was an advanceme in the line of the consideration of principles in natural philosophy. In the *Phaed* 97, Socrates in his discussion of his attempt to discover the meaning of the great nd less, addition, substraction and the reason for the generation and corruption of nything, says:

> Then I heard someone reading, as he said, from a book of Anaxagoras, that mind was the disposer and cause of all, and I was delighted at this notion which appeared quite admirable, and I said to myself: If mind is the disposer mind will dispose all for the best, and put each particular in the best place; and I argued that if anyone desired to find out the cause of the generation or destruction or existence of anything, he must find out what state of being or doing or suffering was best for that thing, and therefore a man had only to consider the best for himself and others, and then he would also know the worse, since the same science comprehended both. And I rejoiced to think that I had found in Anaxagoras a teacher of the causes of existence such as I desired, and I imagined that he would tell me first whether the earth is flat or round; and whichever was true, he would proceed to explain the cause and the necessity of this being so, and then he would teach me the nature of the best and show that this was best; and if he said that the earth was in the centre, he would further explain that this position was the best, and I should be satisfied with the explanation given, and not want any other sort of cause. And I thought that I would then go on and ask him about the sun and moon and stars, and that he would explain to me their comparative swiftness, and their returnings and various states, active and passive, and how all of them were for the best. For I could not imagine that when he spoke of mind as the disposer of them, he would give any other account of their being as they are, except that this was best; and I thought that when he had explained to me in detail the cause of each and the cause of all, he would go on to explain to me what was best for each and what was good for all. These hopes I would not have sold for a large sum of money, and I seized the books and read them as fast as I could in my eagerness to know the better and the worse.
>
> What expectations I had formed, and how grievously was I disappointed! As I proceeded, I found my philosopher altogether forsaking mind or any other principle of order, but

> having recourse to air, and ether, and water, and other eccentricities. I might compare him to a person who began by maintaining generally that mind is the cause of the actions of Socrates, but who, when he endeavoured to explain the causes of my several actions in detail, went on to show that I sit here because my body is made up of bones and muscles....

It is quite obvious that in the very interesting quote from the *Phaedo* th Socrates was seeking in vain for a universal panacea for ignorance of univers causes. He was looking for too much. Actually, Anaxagoras is to be praised fo his finding this principle of causality as well as for his insight into what was t become later the theories of prime matter, of potency and of act.

The mind, as posited by Anaxagoras, had the power of segregating o distinguishing among beings that were, so to speak, "cut off" from the whole. I we push to its logical consequence the theory of the mind, as presented b Anaxagoras, it would have the power of complete segregation. And this would b self-contradictory according to the philosophy of this ancient philosopher, since h said that there was no smallest magnitude but only a smaller. And so, the intellec would be attempting in vain to accomplish what is impossible. Moreover, as w have said, the intellect could not completely separate inseparable accidents from mixture, for this would be to denature these accidents and to take away from then their being, which is an "improper" and unfit action for an intellect.

106. In generation taken strictly, however, there is generation of somethin similar in nature, and when generation is perfect, there is a perfect similitude, as for example, if water would produce something else, that something else woul also be of water. But things can be decomposed into another species, for wheneve there is decomposition there is another species.

)7. The word "from" here poses a problem. We criticize Anaxagoras for ·lieving that a house can be made from a house. We believe that "every agent akes something similar to itself," but this does not mean that the opus must be of e same nature as the agent but that the opus must be the expression of the idea of e agent or of his intention. Here the "from" does not necessarily mean the aterial cause. It does not mean that a house must necessarily be made from a ɔuse. It seems reasonable that Anaxagoras may well have meant by the "from" ›me sort of exemplary cause, for he did posit an intellectual agent.

)8. It is better to have fewer principles than a great number or even an infinite ımber of principles. What can come to be through fewer principles should come › be through the lesser number of principles. The less the number, the more the ıture of principle, for the closer to the nature of beginning or primary element. ccording to this, Empedocles' four elements and even the additional two of ·iendship and strife, are better than the infinitesimal number of particles of ıything and everything of Anaxagoras.

esson Ten

09. Before proceeding it seems that it would be helpful here to discuss the ıeaning of the expression "proceeding from probables." There are two very clear ·xts that give us the meaning of this. They are the *In IV Metaph.*, less. 4, ı. 572–574 and the *In I Post. Anal.*, less. 20, n. 5.

In n. 572 of the *In Metaph.*, St. Thomas tells us that dialectic, which is oncerned with probables, and sophisticates, which is concerned with "apparent ·isdom," are both like and unlike philosophy. And in n. 573, he says:

> They agree in this, viz., that it is proper to the dialectician to consider all things. But this cannot be unless he considered all things to the extent that they agree in some one thing, because one science has one subject and one art has one matter. Since all things agree only in being, it is clear, therefore, that the matter of dialectics is being and what belongs to being. This the philosopher considers. In the same way, sophistics also has some likeness to philosophy. For sophistics is 'seeming' or apparent wisdom, not real (existens) wisdom. But what has the appearance of something must have some likeness to that thing. And, therefore, the philosopher, the dialectician and the sophist must consider the same things.

He continues in n. 574:

> But they differ from one another. The philosopher, indeed, differs from the dialectician in power (virtute), for the consideration of the philosopher is of greater power than that of the dialectician. For the philosopher proceeds demonstratively with regard to the foregoing common things. And it is, therefore, proper to him to have science of the foregoing, and he knows them with certitude, for cognition that is certain, i.e., science, is the effect of demonstration. But with regard to the foregoing, the dialectician proceeds from probable premises, and so, he does not have science but a kind of opinion. And the reason for this is that being is two-fold, viz., being of reason and being of nature. Being of reason, however, is properly said of those intentions which reason comes upon in what it considers, as the intention of genus, species and similar things which, indeed, are not found in the nature of things but which follow upon the consideration of reason. And such being, i.e., being of reason, is properly the subject of logic. But intelligible intentions of this kind are made equivalent to beings of nature in this respect, viz., that all beings of nature come under the consideration of reason. And, therefore, the subject of logic is extended to all things of which being of nature may be predicated. And so, he concludes that the subject of logic is made equivalent to the subject of philosophy, which is being of nature.

> The philosopher, therefore, proceeds from the principles of being of nature to prove what is to be considered with regard to the accidents that are common to such being. The dialectician, however, proceeds from intentions of reason, which are extrinsic to the nature of things, to what should be considered. And, therefore, it is said that dialectics is tentative, because tending is proper to proceeding from extrinsic principles.

In n. 573, we are told that both the philosopher and the dialectician consider eing and what belongs to being and agree in this respect. In n. 574, we are told at the philosopher differs from the dialectician in that he has more power, since is procedure is that of demonstration; and since he proceeds demonstratively, the hilosopher has knowledge that is certain. The dialectician, on the other hand, roceeds from probable principles and, therefore, does not necessarily arrive at nowledge that is certain, but he arrives at a kind of opinion. Then, too, the hilosopher proceeds from intrinsic principles, viz., the principles of being of ature, while the dialectician, like a logician, since dialectics is part of logic, roceeds from extrinsic principles, viz., principles of being of reason or intentions f reason. Dialectics is said to tend towards certain knowledge, since tending is roper to procedure from extrinsic principles, i.e., principles that are extrinsic to e very being of the thing considered. The philosopher achieves certitude, while e dialectician achieves probability; the latter is always on the way to certain nowledge; it progresses towards certitude; there is no arrival, for there is no way y which dialectics as such can either arrive at certitude or know that it has rrived. Yet, dialectics is important to a science, be that science metaphysics or hysics. It is a means by which new paths of investigation may help to make clear hat is under consideration. The use of logical constructs in dialectics may lead a new facet of understanding of reality.

The kind of dialectics described in the above quotations from the *In I Metaph.* is that of reasoning from principles that are purely logical, i.e., principle composed of terms that are based on second intentions. Dialectics of this kind i more like a general logic than a strict dialectics, for dialectics is usually taken i a more strict sense. The dialectics as considered in the *In Metaph.* proceeds, a does all logic, from common principles, logical intentions, beings of reason and n of nature.

The reference cited in the *In Post. Anal.* corroborates what is said in the *I Metaph.* about the fact that both first philosophy and dialectics concern commo things, but the *In Post. Anal.* adds a new facet which gives dialectics a mor restrictive meaning. St. Thomas says:

> Dialectics concerns common things. There is also another science of common things, and this is first philosophy. Its subject is being and it considers what follows upon being, as the proper passions of being.
>
> We should know, however, that dialectics, logic and first philosophy are concerned with common principles in different ways. For first philosophy is concerned with common things since its very consideration is that of the common things themselves, i.e., of being and the part and passions of being. And since reason must consider all that is in things, and since logic considers the operations of reason, logic will be concerned with what is common to all, i.e., with intentions of reason.... The use of demonstrative science is not to proceed from common intentions to show something about things that are the subjects of other sciences. Dialectics, however, does this because the dialectician proceeds from common intentions to what belongs to other sciences, whether that be proper or common, but mostly common.

In other words, the new note added here is that dialectics is not only a mean by which the intellect uses mental constructs to draw conclusions about realit within the science of dialectics, but also to draw conclusions about reality considered in other sciences as well. Dialectics not only argues with regard to the common principles of reason, i.e., second intentions, but it also argues with regard to the common principles of things by using common principles. In this type of dialectics, the terms of the principles are not formed by the mind, although the principles themselves are formed by the mind, for the commonness of universality of the principles is something that depends upon the intellect. The principles used are not, therefore, proper to the science in which the reasoning takes place, but are proper to other sciences.

There is still a third manner of considering dialectics, viz., as proceeding from common principles in the sense of from probable principles. This was briefly mentioned in n. 574 of the *In Metaph.* but is proper to the consideration of the *Topica*. These principles are but opinions and are considered as probable in the sense that they are accepted as certain by all or by people who are intelligent. Yet, they are opinions, for the fear that the opposite is true remains. It is for this reason that if one is said to proceed from this kind of dialectics that he is said to proceed disputatively from probable principles. It is in this sense that Aristotle frequently proceeded, as he is said to have done here at the beginning of less. 10 of the *In Physic*.

The type of dialectics mentioned in n. 573 and for the most part in n. 574 of the *In Metaph.* is a more general kind of logic than that mentioned in the citation from the *In Post. Anal.* and the two types are still somewhat different and more scientific than that studied in the *Topica*. In this third type of dialectics, there is no scientific reasoning, for there is not a strict demonstration. There is but dialectical probability which is the kind of knowledge which *seems* to be certain and which is accepted as such by most people or by intelligent people. The mind gives assent to such knowledge, but it always has fear that the opposite may be true. Just as truth is the adequation of the mind with what is, so is probability the adequation of the mind with what seems or appears to be. The latter is a similitude or likeness to truth. We must also take into consideration here the fact that probability is to be contrasted *not* with truth but with certitude. Probability is a kind of truth, if you wish, but it is not a kind of certitude. A probable judgment is an *opinion*, which is opposed to certitude but which may prove most helpful in the quest for truth. It is for this reason that the dialectics of probable reasoning is so often invoked by St. Thomas.

This kind of dialectics belongs to logic called "logica utens" and not to "logica docens," which merely gives the rules and does not apply them. "Logica utens" or applied or useful logic investigates but it does not resolve its conclusions to first principles. This will become more clear in the Appendix note which follows.

110. Here it seems advisable to consider what is meant by the certitude arrived at by the demonstrative syllogism as compared with probability which is achieved through the use of the dialectical syllogism.

In the *Prior. Anal.*, Bk. I, less. 1, we are told that a syllogism is a "discourse in which, after some things have been set forth, something else different from what was stated necessarily follows because of what was stated." If this is a true definition, then necessity must enter into every kind of syllogism. If the premise of a syllogism is necessarily certain and if the form of the syllogism is correct, then the conclusion is necessarily certain. On the other hand, if the premise is probable and if the form of that syllogism is correct, then the conclusion is necessarily probable. There is, then, a kind of necessity involved in the dialectical syllogism. The form of a syllogism, e.g., a demonstrative or a dialectical syllogism, is determined by the content. The former contains *certain* premises and the latter *probable* premises.

Now, briefly, in the genus of syllogism there are various species, viz., the *demonstrative*, the *dialectical*, the *contentious* or eristic or litigious, and the *paralogism*. Actually, this is not a strict division, for some are syllogisms only figuratively, as we shall see. The species are:

1. *Demonstration*: This is a syllogism that proceeds from premises that are true and first, from indemonstrable propositions or from a self-evident definition.
2. *Dialectical syllogism*: Here the premises are not determined absolutely, for they can be true or false. Since the premises are but probable, the conclusion will be probable. The species is defined by the premises. From premises that are true and first and immediately certain we have demonstration, and from premises that are probable we have the dialectical syllogism. As we have said above, the form of a syllogism is distinct from and, in a way, determined by the content. The kind of knowledge achieved by the dialectical syllogism is opinion, what appears to be true. What is received by most as probable, or as "seeming" truth, is in the province of dialectics and what is received as true, and is really true, is in the province of demonstration.
3. *The contentious, eristic or litigious syllogism*: This syllogism takes as its point of departure opinions which, while appearing to be probable, in reality are not so. This syllogism does not conclude in reality but only apparently and is, therefore, not a true syllogism.
4. *Paralogism*: This syllogism takes as its point of departure premises that are proper to a determinate science and that are neither true nor probable, although they have a vague appearance of probability. This is the worst kind of syllogism.

Now that we have outlined the kinds of syllogisms, we shall consider the distinction between demonstrative certitude, i.e., the certitude engendered by the demonstrative syllogism, and probability, engendered by the dialectical syllogism. In the case of intellectual knowledge, certitude is a property of that knowledge. There is only one kind of certitude and that is subjective certitude. If, however, one asks the grounds of assent, then one can speak of moral, physical or metaphysical certitude, but certitude as such is a quality of knowledge, and here, of course, we are speaking of intellectual and not sense knowledge. The principle of certitude is evidence, intrinsic evidence. In a demonstrative syllogism, the conclusion should be "seen" in the premise and should be able to be resolved to the premise. There is a proportion between the certitude of the conclusion and that of the principle or premise. If there is no deviation from the truth in the premise and if the form of the syllogism is correct, there will be no deviation from the truth in the conclusion. The certitude of the demonstrative syllogism gives rise to science, which is certain knowledge through causes. The effect, the conclusion, is seen in its cause. Opinion, on the other hand, is a probable judgment. There is not certain knowledge; there is no intrinsic evidence; there is conjecture. The premise as well as the conclusion may be true or false.

In the *In VI Ethic.*, less. 3, n. 1143, St. Thomas very aptly says:

> Indeed, we have said above that intellectual virtues are habits by which the soul says what is true. There are, however, *five* of these through which the soul always says what is true either by affirming or by negating. They are *art*, *science*, *prudence*, *wisdom* and *understanding*. Hence, it is clear that those are the five intellectual virtues. But he excludes from this number *suspicion*, which is obtained through some conjectures about some particular facts, and *opinion*, which is obtained through some conjectures about some universal statements. For,

> although through those two the truth may sometimes be said, nevertheless, sometimes they happen to say what is false. The false is the evil of the intellect, just as the true is the good of the intellect. But it is against the notion of virtue that it be a principle of a bad act. And thus, it is obvious that suspicion and opinion cannot be called intellectual virtues.

In the first lesson of *In I Post. Anal.*, St. Thomas, following Aristotle, divides the books of logic according to the three acts of the intellect, viz., apprehension, judgment and reasoning. With regard to the third act of the intellect, reasoning, he speaks of three processes. In n. 5, he says:

> There is, indeed, a process of reasoning that leads to necessity. In this process it is impossible that there be a defect of truth, and the certitude of science is acquired through such a process. There is another process of reasoning in which truth is attained for the most part but not necessarily. The third process is the one in which reason deviates from the truth because of a defect in some principle, which should have been observed in reasoning.

And in n. 6, he clearly distinguishes between science and opinion:

> That part of logic which serves the first process is called *judicative*, since the judgment has the certitude of science. And since we cannot have certain judgment of effects except by resolution to first principles, this part, therefore, is called the *Analytics*, i.e., resolutory. But the certitude of judgment, which is obtained by resolution, is either from the form of the syllogism alone, and to this is ordered the Prior Analytics, which is concerned with the syllogism as such. Or, along with this, the certitude is also from the matter, since the propositions used are 'per se' and necessary. And the *Posterior Analytics*, which is concerned with the demonstrative syllogism, is ordered to the consideration of the certitude from the viewpoint of the matter of the syllogism.

> The second process of reasoning is served by another part of logic. This part is called *inventive* logic, for finding is not always accompanied by certitude.... By means of this process, occasionally, even if we do not obtain science, we still obtain belief or opinion because of the probability of the propositions from which we proceed, since the reasoning inclines wholly to one side of a contradiction, with fear that the other may be true. The *Topics* or *Dialectics* is oriented to considering this, for the dialectical syllogism that proceeds from probable premises is considered by Aristotle in the book of the *Topics*....

Two points should now be clear, viz., that there is a difference and what this difference is between the kind of knowledge obtained in science through the demonstrative syllogism and that obtained in opinion through the dialectical syllogism. Cf. also Appendix note 124.

111. Here we must understand contrariety in a very general sense. In fact, all the ancient philosophers, whether their considerations were of nature or not, stated contrariety in the principles that they posited.

112. The topic of consideration here is that of the contrariety of the principles of mobile beings, and Aristotle quotes the different opinions which they have who posit the contrariety of principles. The second opinion holds that there is only one material mobile principle, in opposition to two, three or four principles. But even this second opinion tries to explain the mobility of mobile being by introducing into one principle a certain contrariety, viz., that of rarity and density. Aristotle speaks here of one *material* principle, because he is speaking about the principles of mobile beings, which are all material.

We should take care not to confound the principles of mobile being with the contrariety of these principles. Contrariety is a note or a quality of those principles, not the principles themselves. It is not the principles that are mobile, but since these principles are principles of mobile being, they must have a certain contrariety.

113. In the *In I De Gen. et Corr.*, less. 22, St. Thomas gives the refutation of the opinion of those who said that indivisible bodies were the principles of things. And, of course, the name and theory of Democritus is evoked. In this lesson, St. Thomas says that Aristotle first shows that indivisible bodies are not principles, and secondly, that they are not moved in a vacuum.

If indivisible bodies are principles, contradictories will be simultaneously true. St. Thomas says:

> If indivisible bodies are principles, they will neither act nor be acted upon mutually. Likewise, if they are principles, they will act and be acted upon mutually: therefore, they will act and not act; they will be acted upon and not acted upon. But this is impossible. With regard to this argument, Aristotle proceeds in this manner: first, he shows for a twofold reason that the principles will not act nor be acted upon. The first reason is that acting and being acted upon takes place in a vacuum, according to these philosophers, but in such atoms there is not a vacuum. It cannot be said, therefore, that these principles act nor can it be said they are acted upon.
>
> The second reason is as follows: Whatever is the cause of an action is either 'hard' or 'soft' or some other property. But no indivisible body is thus. Therefore, neither is it active or passive.
>
> Secondly,...Aristotle shows the opposite, viz., that indivisible bodies may act and be acted upon. He shows this by a threefold argument. The first reason that he gives is this:

According to the foregoing philosophers, indivisible bodies, which are round in figure, are warm. It is necessary, therefore, that an object of another shape be cold, because it is unfitting to place one of the contraries in nature without the other. If, however, there are two qualities in an atom, it is necessary to posit other qualities which follow from the above two qualities, viz., heaviness, durability, lightness, softness, and other similar qualities. When they concede that one atom is heavier than another, as is apparent in a ray of the sun, that one atom has a greater downward tendency than another. If, however, one is heavier than another, one is lighter than another, and one warmer than another.

Since, however, these things are thus, it is impossible that they should not be affected by one another and act upon one another when they come into contact with one another. It is evident that the slightly hot should be affected by one that exceeds it in heat. This would take place not because they are similar in their heat, but insofar as one exceeds the other in heat and is mixed with the one which is cooler.

And St. Thomas continues to make it clear that the position of Leucippus and Democritus with regard to atoms is impossible:

For in atoms either there is figure alone, or there is active and passive quality along with figure. If figure alone is stated in them, then they will be neither active nor passive, because figure is neither active nor passive. Otherwise, mathematical entities would act and be acted upon. But if quality is stated along with figure in the atoms, sometimes there will be one quality in any atom whatsoever, and sometimes there will be many. But if there is one, it will not be proper in any atom whatsoever. This one, indeed, may be hot, and that one cold, and then their nature will not be the same. If, therefore, what differ in nature are divisible, they are divisible into indivisibles. But if many qualities are in one atom, and these qualities, indeed, are active and passive, the qualities will be contraries, because action and passion are in the order of the contraries.... Therefore, contraries are in the same indivisible.

> It also follows that they are in the same thing in the same respect, which is impossible. It also follows that if an atom is chilled, that by this chilling it is warmed. And this is false. And the same is true of other active and passive qualities, as the hard and the soft....

St. Thomas then goes on to give the second principal argument proving that atoms, as stated by the ancients, are not principles. He says that the diverse natures and not the diverse shapes of indivisibles, as Democritus said, should be stated as the principles of action and passion. The reason for this is that what are different in nature and in form act and are acted upon mutually when they approach each other, but what are different in shape only act and affect each other in this way.

> Next, St. Thomas says that Aristotle made it clear that indivisible bodies are not moved in a vacuum, as Democritus said. The argument is as follows: If those indivisibles are moved in a vacuum, we must find out what move them. For either they are moved by themselves, or by another. If they are moved by another, then the indivisible body itself is passive, and, therefore, it is not the first principle of action, but rather, the mover is that principle. But if it moves of itself, either it will be divisible, since, according to one part it moves, and according to another it is moved, or contraries exist in the same thing in the same respect. But both of these are impossible. Thus, not only would the matter of the contraries be one in number, but it would also be one in potency, which is impossible. For if in the matter of the contraries there were unity in potency only, the contraries themselves would have different natures. And thus there would be a multitude of things, but all things would be one, because all would be from the same matter and from the same potency numerically. For since act and potency do not cause difference in species, if there would be one potency, there would be only one species.
>
> But this is to be understood of the potency proximate to contrary forms. And this potency is not one in number, but

> diverse. For there is another remote potency, which is one and the same potency to contraries. And this is prime matter which, as such, is in potency and is its own potency.

Later, assuming now that what Democritus called pores are little vacua, or the interstices between the conjoined indivisible bodies which are the principles of all things, in this same lesson of the *In I De Gen. et Corr.* we learn from St. Thomas that Aristotle said that the existence of these pores is not a necessary condition of action and passion. According to St. Thomas, the first argument given by Aristotle is as follows:

> Whoever say that passions occur from this, that the active moves (movetur) in the pores of the passive,...and that a thing is acted upon when the pores are not filled, must concede that the pores are superfluous. For if some body is acted upon, because the active is in contact with the pores, then that body is acted upon through the act of the active, and not through the pores. Also, if it did not have pores, but is a continuous whole, it will be acted upon in the same way on account of its contact with that which is active. Pores, therefore, are superfluous.

Democritus' theory that all things come to be from conjoined indivisible bodies, which leave pores or little vacua as a result of their being joined or touching each other, seems thoroughly disproved by the fact that indivisible bodies cannot be principles and by the fact that pores are superfluous. It seems that no more need be said to disprove a premise, if the terms of the premise are proved to be irrelevant and impossible in reference to the conclusion.

114. First of all the use of the term "cause" here is equivocal. Some contraries are "caused" from other contraries in the way that the complex is "caused" from the simple. There are contrarieties on various planes and in various genera. Some

contraries may be "caused" from others by subdivision and some may be incidentally "caused" from others. In the *In II De Anima*, less. 21–22, nn. 514–524, St. Thomas speaks of the search for primary contraries in given genera and of how in one genus there can be other contraries besides the first contrariety. He shows how these contraries come to be. He does not speak of any specific mode of cause, however. Actually, this expression "caused" here in the *In Physic.*, could be understood as "come to be."

In n. 514 of the *In II De Anima*, St. Thomas, referring to Aristotle, says:

> He determines with regard to taste, saying that, as in the case of color simple colors are contrary, e.g., white and black, so also in the case of flavors, the simple ones are contrary, e.g., the sweet and the bitter. But the 'adjoining' (habitae) species of taste, i.e., those immediately following upon the simple species, are the succulent which follow upon the sweet, and the salty which follows upon the bitter. The intermediaries of these, however, are the sour, the pungent, the astringent and the acid. The last two are reducible to the same. Almost all the other flavors seem to be reducible to these seven species of flavor.

And he continues in n. 515:

> With regard to these species, it should be noted that although flavors are caused by the hot and the cold, the moist and the dry, and although contraries are those that are the farthest apart, yet contrariety in the species of flavors is not observed according to the maximum differences of hot and cold or of moist and dry but according to order and taste inasmuch as (prout) it is its nature to be affected (immutari) by flavors, either with horror or with delight. There is no need, then, for the sweet or the bitter to be extremes of hot or cold or moist or dry, but that they be extremes in that state wherein they have a reference to the sense of taste....

He is even more to the point in n. 522:

> How it is shown in the *Metaphysics*, Bk. X, that each genus has one primary contrariety. Hence, there must be only one primary contrariety in the case of the object of one sense; and that is why the Philosopher says here that to one sense belongs one contrariety.

But now that he has said that a genus has one primary contrariety, it remains for him to allow for and give the means of obtaining other contraries. He continues in n. 523:

> In one genus, however, there can be several contrarieties besides the primary contrariety, and this either through the mode of subdivision, as in the genus of bodies the first contrariety is that of animate and inanimate. And since animate body is divided into sensible and insensible, and since sensible is further divided into rational and irrational, contrarieties in the genus of bodies are multiplied. Or there are incidentally several contrarities of one genus, as in the genus of body there is the contrariety of white and black and this is true also of all the other accidental qualities of body. And it is in this way that we must understand, with regard to sound and voice, that besides the primary contrariety of low and high, which is essential (per se), there are other contrarieties that are, so to speak, incidental.

115. A procedure to infinity is impossible. If we had to proceed to infinity, we would never find the first contraries, and, consequently, we would never have a real explanation. The first explanation would depend upon the second and so on, in such a way that the first one would never have a real foundation. Infinity cannot be "gone through." In other words, if there were no absolutely first contraries, there would be no contraries at all. If one must find a reason for something, and if he has to proceed to infinity to find this reason, he should stop before he starts.

116. Why cannot action and passion be found in the contingent? There is action and passion only when there is a significant relation the one to the other, and in things that just happen to exist together, there is no significant relation and, therefore, according to St. Thomas, no action and passion. Actually, it is a matter of definition, for significant relation is defined as that which is based upon action and passion.

117. What is the connection between beings that happen to exist together or at the same time and indeterminate beings? A man who, incidentally may be the builder of a house may also be a musician, subway conductor or waiter. There is no determinate connection between the fact that he is a builder or musician or anything else of the kind. And so, from this example, we see that things that happen to exist together (contingentia) are indeterminate. Here, of course, we are taking contingents, i.e., things that happen to exist together, and indeterminates in the very large sense in which they are considered in the *In VI Metaph.*, less. 2, nn. 1184–1185, where St. Thomas gives four examples of the relationship between the contingent or what happens to exist together and the indeterminate. Among these examples, he cites the case of the builder who causes health incidentally, since the builder as builder is not the cause of health.

118. If white came to be from any non-white whatsoever, e.g., sweet, moist, chair, rough, mind etc., and since one could then predicate non-white of almost everything, even of non-being, white, therefore, could be said to come to be from non-being. But such is not the case. The "any non-white" does not mean any non-white whatsoever, but something non-white in the same genus as the white. In the

In X Metaph., less. 6, n. 2038, after explaining that the principle of contrariety is privation and possession, and that this is the first contrariety, St. Thomas says:

> But since privation as such does not receive a more and a less, it can be called perfect privation only if it is in the notion of some nature which has the perfect distance from the habitus. In the same way, not every privation of white is contrary to white, but the privation that is the farthest (magis distans) from white. This privation must be founded on some nature of the same genus and the nature must be farthest from white. And, according to this, we say that black is contrary to white.

119. Non-musical is an infinite term and can be said of all that is not musical. Unmusical, on the other hand, is the contrary of musical in the genus to which musical belongs. Unmusical can be understood as the contrary or privation of musical.

120. A musician can become black, but he does not do so as a musician but as colored. And so, white is destroyed only accidentally in the musician as musician. If a builder who is white became black and in becoming black would become musical, we could say that white was only accidentally corrupted into musical.

121. This is a very important point. A being in the process of change "contains" the *non-being* towards which it is. And that which a term is to become "contains" the non-being of the *other* term of the process.

In order that a living tree be corrupted into a dead tree, the living tree would have to have contained in it the privation or non-being of that into which it is corrupted, viz., dead tree, which is non-treeness, since it is no longer a tree but merely a completely dead tree. And thus, the non-being of what a thing is to become must be contained in the other term of the process.

It is impossible that there be such a thing as a dead tree, unless there were beforehand a living tree. A dead tree contains in itself in some way the non-being of a living tree in the sense that the dead tree is the corruption of the being of the living tree. And thus, the non-being of the living tree is contained in the other term of the process, viz., the dead tree.

122. In the *Cont. Gent.*, Bk. II, chap. 64, where it is proved that the soul is not a harmony, we read: "Harmony is said in two ways: in one way as the *composition itself*, and in another way as the *mode of composition*."

And in the *In I Ethic.*, less. 1, n. 5, after he states that man since he is a social animal, belongs both to a domestic and a civil group, St. Thomas says:

> We should know, however, that this whole, which is the civil group, or the domestic family, has *unity only of order* and, accordingly, is not something absolutely one. And, for this reason, a part of this whole can have an operation which is not the operation of the whole, as a soldier in the army has an operation which is not the operation of the whole army. Yet, the whole itself also has an operation which is not proper to any one of the parts but to the whole, e.g., the battle of the whole army. And the piloting of a ship is the operation of a group piloting the ship. There is, however, a whole, which has unity not only of order but of *composition*, either of *colligation* or even of *continuity*, and according to this unity something is one absolutely. And, for this reason, there is no operation of the part which is not the operation of the whole, for in what is continuous the movement of the whole is the same movement as that of the part. And similarly in composites or in things bound together, the operation of the part is principally that of the whole.

And so, such a consideration both of the whole and of its part must belong to the same science. But it does not belong to the same science to consider the whole which has unity only of order, and its parts.

The division of unity may be expressed briefly by the following:

1. *Unity of order*, e.g., an army. Here there is only a unity of order. There can be operation both of the part and of the whole.
2. *Unity of composition*, e.g., a house. Here there is harmony of parts that are entirely physical.
   a. *Unity of colligation*, e.g., a train caboose or isenglass. Here the operation of a part is mainly that of the whole.
   b. *Unity of continuity*, e.g., an organism. There is no operation of the part that is not that of the whole.

123. The predecessors of Aristotle did not show how principles had to be contraries but they did it only *as if* they were "compelled by truth itself." They were not moved by some argument to a scientific grasp of the fact that principles were contraries. Rather they had an intuitive grasp of this fact. They had this intuitive grasp, as if they were swept into it by a compelling truth.

124. Those who without good reason proposed first principles as contraries cannot have intrinsic certitude of the first principles but only opinion. This latter may suffice to incline the intellect to truth, but it is not sufficient for the attainment of certitude. There is no other way of having certitude in this order of natural philosophy except by experience and demonstration, for in the natural order the intellect can never have certitude without evidence either of fact, and here there is the certitude of sense, or the evidence of principle and conclusion, and here there

s demonstrative or intellectual certitude. There is, however, a kind of inclination of the intellect to truth by reason of opinion.

Actually, the distinction between certain knowledge and opinion is one which, lthough it belongs to the realm of logic, is pertinent to all the sciences, for logic nay be considered as the method of science. In the *In III De Anima*, less. 4, . 632, St. Thomas says:

> When we know (intelligimus) intelligible objects, we assert that they are such. But when we form opinions, we say that they seem to be such and such or that they appear to us to be such and such; for just as to know (intelligere) requires sense, so also to have an opinion requires imagination.

Since opinion proceeds from probable arguments, it is most interesting to ote that here in n. 1 of this lesson of the *In Physic.*, St. Thomas says that "first e investigates it by the method of disputation by proceeding from probables." In ther words, the method of investigation is the same as the one that the ancient hilosophers used in drawing their conclusions about first principles and con-rariety. This method, of course, was that of proceeding from probables, the nethod of opinion, and not the method of certitude obtained by demonstrative easoning.

25. Some things are more known intellectually and others more known sensibly. uch notions as those of harmony and discord are more abstract in themselves than re such notions as the humid and the dry. Since the former more general notions re more abstract, they are more known according to reason.

Moreover, the humid and dry are attributes of sensible matter, while discord nd harmony, even and odd, comprise an order established by the mind. These notions are more universal and abstract in that they may be considered in themselves without application to sensible matter or they may be superimposed upon such matter. The humid and the dry, on the other hand, since they are applicable only to sensible matter, are less universal and better known to the senses.

Moreover, discord and harmony are respectively the denial and affirmation of order. But order is only perceived by an intellect. Things may have order in objective reality but this order is only known by the intellect. Order always bespeaks reason; it is reasoned out, so to speak.

126. Pythagoras advanced good reasons for saying that the substance of all things was number. First, all corporeal substances were measurable and measured by some number. Burnet, *op. cit.*, p. 52 tells us that "the early Pythagoreans represented numbers and explained their properties by means of dots arranged in certain 'figures'...or patterns." These figures are not the Arabic figures, but geometric forms, as squares, triangles or oblong forms. The dots in these forms were called "terms" or "limits," while the space around them was called the "field."

The later Pythagoreans had to contend with the doctrine of Empedocles that was popular at that time. They said that the four elements of Empedocles were "'figures' or, in other words, that they were made up of particles which had the shapes of the regular solids," according to Burnet, *ibid.*, pp. 88–89. This introduces the second reason why the Pythagoreans said that the substance of all things was numbers, viz., all corporeal substances were divisible into elements and he said that the difference between one element and another is a difference in number. If the elements were figures and if figures are formed by the addition and arrangement of dots than corporeal substances that are made up of elements are divisible into these elements. Since elements are figures, corporeal substances are divisible into figures. And since the elements differ by figure or particles having a different shape, they differ in number, for by number the Pythagoreans meant figure.

Elements are figures, and so since elements are the substance of all things, the substance of all things is figures or number. And so the things themselves, i.e., the things composed of this substance or a combination of these elements which are numbers, are in a sense numbers. However, Burnet, *ibid.*, p. 89, says:

> The later Pythagoreans appear to have said that things were *like* numbers rather than that they actually were numbers, and here we shall probably be right in tracing the effort to Zeno's criticism. Aristotle quotes the doctrine in both forms, and he hardly seems to be conscious of any great difference between them.... What the later Pythagoreans probably meant by saying things were 'like numbers' instead of saying that they actually were numbers...must have been something like this. For the construction of the elements we require, not merely groups of 'units having position,' but plane surfaces limited by lines and capable in turn of forming the limits of solids. Now Zeno had shown that lines cannot be built up out of points or units, and therefore the elementary triangles out of which the 'figures' are constructed cannot be identical with triangular numbers.... In particular, the isosceles right-angled triangle is of fundamental importance in the construction of the regular solids, and it cannot be represented by any arrangement of 'pebbles,' seeing that its hypotenuse is incommensurable with its other two sides. It only remains for us to say, then, that the triangles of which the elements are ultimately composed are 'likenesses' or 'limitations' of the triangular numbers.

We have spoken of number as the element of all things, but of the element of number we have said only that it is figure or shape. Burnet, *ibid.*, pp. 320–321, explains the nature of number in another way. He says:

> The Pythagoreans had regarded the Limit...and the Unlimited...or Continuous as the elements of number, and therefore as the elements of things. Plato substituted for these the One and the dyad of the Great-and-Small. The only difference, according to Aristotle, is that the Pythagorean Unlimited was single, whereas Plato regarded the 'matter' of numbers, and therefore of things as dual in character. It also follows, as Aristotle points out elsewhere, from Plato's separation of numbers and things that there will be what he calls 'matter' in the numbers as well as in things. This is called the Indeterminate dyad...to distinguish it from the Determinate dyad, which is the number two. From this dyad the numbers are generated as from a sort of matrix....
>
> Now it is at least clear that the term Indeterminate Dyad is a new name for Continuity, and it expresses more clearly than the old term Unlimited its twofold nature. It not only admits of infinite 'increase'...but also of infinite 'diminution'.... That is why it is also called the Great-and-Small. The new idea which Plato intended to express was that of the infinitesimal.... The introduction of this conception involves an entirely new view of number.

For the Pythagoreans, then, the unlimited, i.e., the infinite, was the even because the even was divisible. So also for Plato, the unlimited, i.e., the Great-and-Small was what had the capacity of being limited by form. The odd which was not capable of division, and which was the definite, the divided but not the divisible was the form. It was the limited and not the unlimited. Both Plato and Pythagoras stated the unlimited to be on the side of matter and the limited or limiting to be on the side of form.

Yet, we should add that being for Pythagoreans and being for Plato was different. Burnet, *ibid.*, p. 331, says:

> The Pythagorean doctrine simply identified the Form with being and the Unlimited with becoming, but Plato distinctly

> states that the Mixture alone is truly 'being.' The process of mixing is indeed a 'becoming'...but it is a becoming which has being for its result...and the mixture itself is being, though a being which has become.

127. "Discord" and "harmony" are the effect of the principles "friendship" and "strife" which were considered in nn. 13 and 57 of the text of the *In Physic.* In less. 2, of the *In I De Gen. et Corr.*, St. Thomas says that Aristotle, speaking of one subject and many contraries as principles of being, gives two arguments against Empedocles.

> With regard to the first of these, he says that Empedocles seems to say things contrary not only to what is sensibly apparent, in which we see that water comes to be from air, and from air fire, but he also seems to contradict himself. For, on the one hand, he says that no element is generated from another but that all other corporeal elements are composed from them. On the other hand, he says that before this world was generated, every nature of things except strife happened to be joined into one through friendship. And he says that again any one of the elements, and also any one of the other bodies will be segregated through strife, the segregator of things. Hence, it is clear that certain differences and attributes of diverse elements came to be through strife, and that from that one first principle, viz., strife, one element would be water and another would be fire.
>
> And he gives examples of differences and attributes, e.g., he says that the sun, i.e., fire, is white, hot, and light, but the earth heavy and weighty. And thus, it is evident that these differences repeatedly come to and go from the elements. But whatever comes and goes can be taken away. Therefore, since differences are removable, insofar as they are generated repeatedly, it is obvious that when differences of this nature are removed, it is necessary both that water come to be from earth and earth from water. And likewise, any other element must come from another.

> And this must happen not only then, viz., in the beginning, but also now: and this must take place through the transmutation of passions. And that such transmutation can come to be, he proves in two ways. First, he proves it from the nature of the passions themselves, because from those which Empedocles speaks of, it follows that transmutation can come to it again and again, e.g., through strife, which does the segregating and it can be separated again from the elements, viz., through the coming of friendship.

After proving through *the nature of the passions* that the transmutation of th passions of the elements can take place, St. Thomas tells us that Aristotle als proved this from the *cause of those passions*. In the same lesson, St. Thoma continues:

> Since even now strife and friendship are contrary to one another, so also were they in the beginning. Therefore, even now elements can be transmuted according to differences and passions. And, therefore, at that time, viz., in the beginning of the world, the elements were generated from one thing and then these differences came to the elements. For it cannot be said that fire, earth, and water actually existing form one totality.
>
> The second argument he treats later.... And he says that it is uncertain whether Empedocles should have stated one principle or many, even though he actually posited many, viz., fire, earth, water, and others which co-exist with them. But he says that this is uncertain because from the fact that he stated one principle from which, as from matter, come fire, earth and water, through a certain transmutation brought about through strife, it seems that there is one element, but insofar as that one comes to be from the composition of elements coming together into one through friendship (but those, viz., the four elements, insofar as they come to be from that one element through a certain breaking up through the workings of strife), it seems rather that those four are the elements and principles, and what is prior in nature.

> And Empedocles gave more attention to this, when he stated that things come to be through coming together and separation. Aristotle, however, in the preceding arguments proves that it is necessary for elements to come to be not through separation alone, but through a certain transmutation that comes to the differences of the elements. From this, there follows the contrary of what Empedocles maintained, viz., that that one element is more a principle.

St. Thomas even more specifically and more forcefully criticizes the theory of Empedocles which claims discord and harmony, or friendship as principles of generation. This criticism is in the *In II De Gen. et Corr.*, less. 7, where he says:

> For three reasons Aristotle destroys the opinion of Empedocles with regard to moving principles. The first of these arguments he states when he says that Empedocles treated motion insufficiently. For it is not enough simply to say that friendship and strife are moving principles....
>
> The second reason...is as follows: Natural bodies are moved violently and not naturally and also naturally, e.g., fire moves upward not by violence but by nature. But it moved downward violently and what is violent is contrary to what is natural. For we see that bodies are moved by violence; therefore, they are moved according to nature, for the becoming of contraries is related to one and the same thing.
>
> Let us investigate from the theory of Empedocles whether the motion of friendship is natural or violent: for that earth is moved upward just as that fire is moved downward is against nature. But what is against nature is likened to a certain segregation. On the other hand, the elements will thus be brought together because unless fire descends and earth ascends, the elements will not be brought together. Since, therefore, bringing together is the effect of friendship, according to Empedocles the motion of friendship is violent and like segregation, and strife is more a cause of natural motion than is friendship, which is impossible according to him. And, therefore, it is impossible to state that friendship and strife are

> moving principles. But if these are not moving principles, and since there are no other principles according to Empedocles, there is no motion of natural bodies, nor is there any rest, for possession and privation would have becoming with regard to the same thing. This, however, is obviously false.

It would seem, then, according to Aristotle and St. Thomas, that neither discord and harmony nor friendship and strife can be posited as principles of the generation and corruption of things.

128. Cf. Appendix note 125.

129. "Better" and "worse" are words which have a value significance for us and are taken analogously here with reference to principles. St. Thomas means that some principles are better or worse as principles than others.

130. In the *In X Metaph.*, less. 6, n. 2036, St. Thomas says:

> The principle of contrariety is privation and possession; and he says that the first contrariety is privation and possession, because privation and possession are included in every contrariety.

And in n. 2037, he clarifies a doubt that might arise:

> But lest someone might believe that to be opposed in the opposition of privation and possession and to be opposed in an opposition of contrariety are the same thing, he adds that not every privation is that of contraries, because privation, as has been considered in the foregoing, is said in many ways. For sometimes, if in some way something does not have what it should have by nature, there is said to be privation. But this kind of privation is not that of contraries, because this kind of

privation does not state some nature opposed to possession, although it may suppose a determined subject. But privation is said to be contrary when it is a perfect privation.

131. "All assume principles that are better known" in the sense that they all agree that what is a principle should be what is better known.

Lesson Eleven

132. Substance has one first contrariety, but this does not prevent there being other contrarieties under the heading of the same genus. In the *Categoriae*, chap. 5, 11. 4 a 10—4 b 20, Aristotle says:

> The most distinctive mark of substance appears to be that, while remaining numerically one and the same, it is capable of admitting contrary qualities. From among things other than substance, we should find ourselves unable to bring forward any which possessed this mark. Thus, one and the same colour cannot be white and black. Nor can the same one action be good or bad: this law holds good with everything that is not substance. But one and the self-same substance, while retaining its identity, is yet capable of admitting contrary qualities.... It is by themselves changing that substances admit contrary qualities....
>
> It is by reason of the modification which takes place within the substance itself that a substance is said to be capable of admitting contrary qualities; for a substance admits within itself either disease or health, whiteness or blackness. It is in this sense that it is said to be capable of admitting contrary qualities.
>
> To sum up, it is a distinctive mark of substance, that, while remaining numerically one and the same, it is capable of admitting contrary qualities, the modification taking place through a change in the substance itself.

Although a bit repetitious, the above makes it clear that the capacity of substance to have contrary qualities is by reason of something that takes place within substance. Within each substance there are many contrarieties, but there is one first contrariety which serves as a basis for the other contrarities.

133. If we state contraries alone to be principles, this contrariety will inevitably be among accidents, because substances are not contraries. Contrariety exists between the opposites in the same genus. Substance has no contrary. In the *Categoriae*, chap. 5, 11. 3 b 24—3 b 31, Aristotle says:

> Another mark of substance is that is has no contrary. What could be the contrary of any primary substance, such as the individual man or animal? It has none. Nor can the species or the genus have a contrary. Yet this characteristic is not peculiar to substance, but is true of many other things, such as quantity. There is nothing that forms the contrary of 'two cubits long' or of 'three cubits long', or of 'ten', or of any such term. A man may contend that 'much' is the contrary of 'little', or 'great' of 'small', but of definite quantitative terms no contrary exists.

Moreover, contrariety is the opposition of what is farthest apart in the same genus. But there is no degree of variation in substance. Aristotle continues in the same chap., 11. 3 b 32—4 a 9:

> Substance, again, does not appear to admit of variation of degree. I do not mean by this that one substance cannot be more or less truly substance than another, for it has already been stated that this is the case; but that no single substance admits of varying degrees within itself. For instance, one particular substance, 'man', cannot be more or less man either than himself at some other time or than some other man. One man cannot be more man than another, as that which is white may be

> more or less white than some other white object, or as that which is beautiful may be more or less beautiful than some other beautiful object. The same quality, moreover, is said to subsist in a thing in varying degrees at different times. A body, being white, is said to be whiter at one time than it was before, or, being warm, is said to be warmer or less warm than at some other time. But substance is not said to be more or less than which it is; a man is not more truly a man at one time than he was before, nor is anything, if it is substance, more or less what it is. Substance, then, does not admit of variation of degree.

It seems quite obvious, then, that there is no contrary of substance. Contraries are accidents. Accidents are predicated of a subject. This subject is prior to the accidents that are predicated of it. If a first principle were a contrary, it would be an accident predicated of a subject. But a subject, by its nature, is prior to an accident. Thus, the first principle would be predicated of what is naturally prior to the first principle. Since what is first is not predicated of what is prior to it, this theory of a first principle's being a contrary is impossible. An accident would be prior to its subject.

134. If we took two probable solutions and put them together, we still would not have certitude. No amount of probability produces certitude.

135. The subject cannot have a contrary, for then these two contraries would in turn need a subject or substrate. Since they are accidents, contraries need a subject to which they can be accidents.

In the *In II De Anima*, less. 22, n. 524, St. Thomas says:

> In the genus of what can be touched there are many first contrarieties strictly speaking (per se) but all these contrarities

> are in a way reduced to one subject and in a way not. *In one way*, the very genus of contrariety, which is compared to the different contraries, as potency to act, can be taken as the subject of contrariety. *In another way*, substance, which is the subject of the genus of which there are contrarieties, can be taken as the subject of contrarieties, as if we were to say that a colored body is the subject of white and of black. If, therefore, we were to say of a subject that it is a genus, it is clear that it is not the same subject of all the qualities of things that can be touched (tangibilium). But if we were to speak of a subject which is substance, there is one subject of all of these, viz., body, belonging to the make-up (consistentiam) of animal....

There may be a subject of contraries, but not a contrary of a subject. Contraries are *in* a subject, but a subject could not have a contrary, because a subject would have to be in another subject. This would denature the meaning o subject. The requisites for a contrary are stated in the *In V Metaph.*, less. 12 n. 926, where St. Thomas says that contraries, in the proper meaning of contraries agree in three things:

> Contraries, indeed, agree in three things: viz., in the same genus, and in the same subject and in the same capacity. And from these three conditions he elucidates those things that are truly contraries. He says that of those things that are in the same genus, those that differ the most are called contraries, e.g., black and white in the genus of color. And again, those that differ the most in the same existing thing that is capable of receiving them (susceptibili existentia), e.g., healthy and sick in an animal, are called contraries. And again, those are called contraries that differ the most, that are contained in the same capacity, e.g., congruous and incongruous in grammar (grammatica)....

From this we see that contraries are not subjects but require a subject and in fact, must be in the same subject. If a subject were a contrary, this would mear that the whole subject would have to be a contrary. But according to the meaning of a contrary, the contrary of subject would have to be in the subject, which is a contrary. This, of course, is impossible. There can be contraries in a subject, but these contraries are not the contrary of the subject.

136. They were right to the extent that they posited fire as the first principle because of its subtlety, but wrong to the extent that fire has the very sensible quality of warmth. This "sensible" aspect of the principle fire makes the principle fire to be less simple and, therefore, less prior. But the ancients considered only the subtlety and not the "sensibility" of fire.

In the *In I Metaph.*, less. 12, n. 186, St. Thomas says that, for the ancients who held fire to be a principle, whatever was generated through concretion came to be from this first principle. He says:

> That from which other things come to be through concretion is what is the most subtle among bodies, i.e., what has the least amount of parts. And what was more simple seems to fulfill this requirement. Hence, if the simple is prior to what is composed, it seems that this is first....

And he adds that "whoever posited fire as a principle posited it as the first principle because it is the subtlest of bodies...."

Also in the same work, Bk. XI, less. 1, nn. 2171–2172, St. Thomas says that the more simple, because it is less divisible, seems more to be a principle than the less simple.

In the *In III De Caelo*, less. 9, St. Thomas says that after Aristotle showed that elements are not infinite, he went on to make it clear that they are not just one either, although they must be finite. At the beginning of the lesson, St. Thoma[s] says:

> In the first part Aristotle says that some of the ancients posited only one element of bodies: some said that it was water, as Thales...because the sperms of animals and the food of animals and plants seemed to be moist. Others, however, said that it was air, as Anaximander and Diogenes, because they saw that it was easily alterable to anything at all. Others even said that it was something intermediate between these, something more subtle than water and grosser than air that was infinite and contained all the heavens and all bodies universally, as did Anaximander. Others, however, said that it was fire...as Heraclitus, the Ephesian, because they saw that it was the most active element of all.... They say that fire is the most subtle of all bodies and, therefore, will be the prime nature of all things.

Later on in the same lesson, we are told what they meant by "subtle," viz. "something rare," i.e., refined, of few parts, of little bulk.

The remainder of the lesson is devoted to disproving the theory that fire o[r] any other *one* element is *the* element of bodies.

137. Cf. Appendix notes 85 and 127.

138. Even Thales posited three principles when he said that there was water and the rarefaction and condensation in water. Here he had two contrary accidents and one substance. There is a multiplicity that comes from the form and a multiplicity that comes from the matter. We cannot posit matter as a principle of unity only or as a principle of multiplicity only.

139. He is proceeding dialectically here. Even if there were just the four contraries, for each of the pairs there would have to be one subject.

140. Here he is considering things in general and in a confused manner. And so, in this second argument, he reverts to substance and not to any other genus.

141. Contrariety is an ontological distinction, while that of non-contradiction is logical. In the order of contrariety there is no intermediary. There is no way for one to become the other. Let us take, for example, the contraries of animate and inanimate, sensitive and not sensitive, rational and irrational, that are found in the substance man. These contraries are related as prior and posterior. Animate and inanimate is a contrariety that is first in the sense of being the lowest of the foregoing contrarieties. To animate may be applied the contrariety of sensitive or not sensitive; to sensitive may be applied the contrariety of rational or irrational. These contraries are not substances, but are hierarchically formal differences found in the genus of substance.

Lesson Twelve

142. In the *Summa Theol.*, IIIa, qu. 17, art. 1, ans. 7, St. Thomas says:

> It should be known that *other* implies the difference of accident and, therefore, the difference of accident is sufficient for something to be called absolutely *other*. But *another* implies the difference of substance. But substance is not only called nature, but supposite, as is said in *Metaphysics* V. And, therefore, the difference of nature is not sufficient for something to be called absolutely another, unless there be difference in supposite. But the difference of nature makes another in a

> certain respect, i.e., according to nature, if there is no difference of the supposite.

It is important to note here the distinction between "other," which in Latin is "alterum," and "another," which is "aliud."

143. The sum of all this seems simply to say that in becoming, both the terminus from which and the terminus to which may be either simple or composite terms. And becoming is said to be either simple or composite according to the simplicity or composition of the terminations. The subject, e.g., non-musical man may be composite, and there is said to be becoming as a composite from the terminus 'from which.' Yet, the predicate may be a simple one, e.g., musical, and here the term is, therefore, said to become as simple. In other words, the kind of becoming predicated of a terminus depends upon the kind of terminus, whether simple or composite.

144. It is better to say that "this becomes this," e.g., that man becomes musical, for this is properly said of a subject that remains, than "this comes to be from this," e.g., musical comes to be from man, for this is properly said when the subject does not remain. The whole point of this lesson is to show that in every becoming a subject always remains.

145. In the *In VII Metaph.*, less. 6, n. 1414, St. Thomas says:

> Indeed, that from which something comes to be as from matter is sometimes predicated not abstractly but denominatively. For some things are said not to be *that*, i.e., matter, but *of the nature of that*, as a statue is not called *a stone*, but is said to be *of stone*. But a man, who is convalescing, *is not called*

> *that from which*, i.e., he does not receive the predication of that from which he is said to come to be, for convalescence comes to be from illness. Nor is it said that the convalescent is ill.

A man called a convalescent is a convalescent by something that is going on within him. On the other hand, a statue is called *of stone* by denominative predication and not by extrinsic denomination. In denominative predication, we denominate a thing on the basis of the material of which it is made, e.g., a table is of wood or a statue is of stone. Nor is the statue called *a stone* in a predication by identity.

And in n. 1415, St. Thomas continues:

> There is a cause of predication of this nature, since a thing is said to come to be from another in two ways, viz., from privation, and from a subject, which is called matter, as it is said that a man becomes healthy, and that the laborer becomes healthy. But it is said that a thing comes to be from privation rather than from a subject, as someone is said to become healthy from working rather than from the fact of his nature as man. But we say that this comes to be this in subject rather than in privation, since we more properly say that man becomes healthy rather than that the laborer becomes healthy. And, therefore, whoever is healthy is not called a laborer, but rather he is called man. And conversely, a man is called healthy. In this way, therefore, what comes to be is predicated of the subject but not of the privation.

The privation is understood and is not that to which the becoming is predicated. For example, in the case of man's becoming healthy, health is predicated of the man and not to the privation of health in him. So also, when man becomes cold, cold is not predicated of the privation of cold, but of man. Yet, it is by reason of the understood privation that there can be becoming and that becoming can be predicated of what becomes.

146. An historico-linguistic comment might not be out of order here. The discussion with regard to *this coming to be this* and *this coming to be from this* has significance only in the framework of a particular way of speaking. We say that a statue is bronze and we know that bronze is an adjective. They, however, use the term *bronze* as a noun but they use *brazen* as an adjective. And for them, as is indicated in the passage in this text of the *In Physic.*, bronze meant "unshaped," and hence, they correctly speak of the subject, e.g., statue or man and its opposite, e.g., unshaped or unmusical.

147. Here in *In VII Metaph.*, less. 6, St. Thomas systematically proves that "there must be a subject in every natural becoming." Referring to Aristotle, he says in n. 1381:

> Here he intends to show that the quiddities and forms existing in those sensibles are not generated from some forms apart from matter, but from forms which are in matter. And this will be one of the ways by which the position of Plato is destroyed. Plato posited separated species which he stated as necessary for this, viz., that through them there might be a science of those sensible things, and for this, that sensible things might exist through participation in these, and also for this, that there might be principles in the generation of sensible things. In the preceding chapter, however, he already showed that separate species are not necessary for the science of sensible things nor for their being, since the essence of a sensible thing which exists in a sensible thing suffices for this and is identical with it.

And in n. 1386, he says:

> ...in natural generation that *from which* what is generated come to be is called matter. But that *by which* is generated some one

> of those things, which are according to nature, is called the agent.

And after discussing matter, the agent, and the composite, St. Thomas says that matter and the agent are as principles of generation, and that the composite is as a termination of generation. In 1388, he very explicitly tells us that there is a subject in every natural becoming, and that this subject is matter. Referring to Aristotle, St. Thomas says:

> He proves that one of the three, viz., the principle out of which, is found in every generation, and not only in every generation, but also in artificial generation (for it is obvious with regard to the other two principles). He says that everything that comes to be either in nature or in art has matter from which it comes to be. For everything that is generated either in art or in nature is able to be and not to be. For, since generation is change from non-being to being, indeed, what is generated must at one time be and at another time not be. This would not be the case, if it were not possible to be and not to be. But in any one thing, what is in potency to be and not to be is matter, for it is in potency to forms through which things have being and to privations through which things have non-being.... It remains, therefore, that matter is necessary in every generation.

Lesson Thirteen

148. What has been discussed up to this point remains very indeterminate in comparison with what is to follow.

149. There seems to be a difficulty here, for he uses subject and form, which are distinct, as the components of a definition. One might ask how things into which the definition of something is resolved can be distinct. The answer is that they

must be distinct as far as the meaning that each signifies is concerned, i.e., they are different formalities making up one definition.

150. The way in which privation is an accidental principle seems clearly explained by St. Thomas in *De Princ. Nat.*, chap. 2, *op. cit.*, pp. 82–85, where he says:

> There are...three principles of nature, i.e., matter, form and privation. One of these, viz., the form, is that to which there is generation, but the other two are on the side of that *from which* there is generation. And so, matter and privation are the same in subject, but they differ in meaning. For the same thing that is bronze, is unshaped before it has the form, but it is called bronze by one meaning and it is called unshaped by another meaning. And so, privation is not called a substantial principle, but an accidental one because it coincides with matter, just as we say that the doctor incidentally builds, for the doctor does not build as a doctor, but as a builder, and this coincides with doctor in one subject. But there are two kinds of accidents: viz., *necessary*, which is not separated from the thing, e.g., risible from man, and *not necessary*, which is separated, e.g., white from man. And so, granted that privation is an accidental principle, it does not follow that it is not necessary for generation, because matter is never devoid of privation, for to the extent that it has one form, it has the privation of another, and conversely, as in the case of fire there is the privation of air and in air there is the privation of fire.

And he continues:

> And we should also know that, since generation is from non-being, we do not say that negation is a principle, but rather that privation is, because negation does not determine a subject for itself. Indeed, not to see can be said even of non-beings, as 'a chimaera does not see.' And it can also be said of beings which by nature are not meant to have sight, e.g., a stone. But privation is said only of a determinate subject, viz., one which

is meant to have the habitus by nature, as blindness is said only of what is meant to see by nature.

And because generation does not come to be from absolute non-being, but from non-being which is in some subject, and not in any subject whatsoever but in a determinate one,—for fire does not come to be from any non-fire whatsoever but from such non-fire that is meant to have the form of fire by nature—therefore, it is said that privation is a principle. But it is different from the others in this, viz., that the others are principles both in being and in becoming. For an idol to be made there must be bronze and lastly there must be the figure of an idol. Moreover, when the idol now exists, these two things must also exist. But privation is a principle in becoming and not in being, because, while the idol is in process, it must not be an idol. For, if it were, it would not be in process, because what is in process exists only in successive stages, as time and motion. But by the fact that the idol now exists, there is no privation of the idol there, because affirmation and negation do not exist at the same time, and neither do privation and habitus. In this way, privation is an accidental principle... but the other two are substantial principles.

From what has been said matter is obviously different in meaning from form and from privation. For matter is that in which form and privation are understood, as figure and unshaped are understood in bronze. Sometimes, indeed, matter is named with the privation, and sometimes without it, as bronze, since it is the matter of the idol, does not imply privation, because undisposed or unshaped are not understood by the word 'bronze.' But flour, since it is matter with reference to bread, implies in it the privation of the form of bread, because the word 'flour' signifies indisposition or a lack of ordering that is opposed to the form of bread. And because in generation, matter or the subject remains and privation does not, nor does the composite of matter and privation, therefore, matter that does not imply privation remains but what implies privation does not remain.

In the above quotation, St. Thomas said that to be of bronze and to be unshaped are two distinct notions, and that there is always privation, since there is no becoming without it. It is for this reason that privation is a principle, even thought it is only an accidental one.

And as St. Thomas said, there are two kinds of accidents. There are necessary ones, e.g., risible for man, and this is inseparable from man. There are also accidents that are not necessary, e.g., white for man. This is not an inseparable accident, for there are men of other colors. In the case of privation, we are not employing the category of accidents that are not necessary, for privation is a necessary and inseparable accidental principle of becoming, but it is not a necessary accident as risible in the case of man. So, we may ask why he uses privation as a necessary accident, since it does not fall under either category, viz., the necessary accident or the accident that is not necessary. The answer is that there is still a third mode of accident, and this is the "per accidens necessarium." The following are the three kinds of modes of accidents mentioned:

1. *The necessary accident*: e.g., risible for man: there is an essential or "per se" connection.
2. *The "per accidens" necessary accident*: e.g., privation in becoming: this is an accidental principle.
3. *The contingent or not necessary accident*: e.g., white for man: there is no essential or "per se" connection.

151. Privation must be an accidental principle in the very becoming of a given mobile being. Privation is not a cause of the being of things any more than musical as musical is the cause of the act of building. With reference to a cause or in the strict sense, privation has no such causal influence. In the case of the becoming of a thing, it is as a *terminus from which*.

152. We must not confuse privation and matter. They are the same in subject, but they are not formally the same. There is a distinction in the notions that each signifies. Privation here is not an indefinite privation of anything whatsoever, but is privation in a definite subject. Privation, therefore, is very determined and is a relative negation, because it is in a subject determined not only in species but even as to the individual of the species.

Privation is, so to speak, a concomitant of that in which it is. No matter what form a subject may have, it still has the privation of other forms. Privation is an extrinsic determination of a subject in that it is a negation of something in a given subject. Privation is more determinate in the sense of being more definite than prime matter. The latter is uninformed by any form that is definite. The former is the privation of a given or determined form. Yet, prime matter, although it is indeterminately oriented to form, can only be informed by one substantial form at one time.

Privation is always the privation of something in a given subject that already exists. The matter of a subject is either individual sensible, common sensible or intelligible. As the matter of a subject, matter is no longer called prime matter. It is called by a more specific name. In short, privation is the privation of something specific in a determinate subject, while prime matter is a more vague indetermination to form, but when it is informed by form, it must be informed by one substantial form and only one.

153. There is no contrariety in substantial generation since this is a procedure from non-being to being, from the negative to the positive. The two opposites are not both positive, as in the case of the opposites that are contraries.

154. Since prime matter is the first subject and, since a thing is known only by its form, prime matter is not known in itself, but it is known analogously, i.e., by comparison to artificial things that are made from natural things, as in the proportion that prime matter is to its form as matter of art, e.g., bronze, is to the work of art, e.g., a statue.

There are many texts on the consideration of prime matter as the first subject. We shall cite two of these, viz., the *De Princ. Nat.* and the *In VII Metaph.*, less. 2, nn. 1277 and 1276 of St. Thomas. In the former, *op. cit.*, pp. 85–86, we are told that prime matter is so called, because it is confusion or matter that is not yet put in order, e.g., into the order of boards and bricks. And we are also told that there is no other matter prior to it. It is the matter that is first subjected to the meaning of matter and, since knowledge of a thing is acquired through the form of that thing, matter can be known only by comparison to form. St. Thomas says:

> We must know that some matter has the composition of form, e.g., bronze, although it is matter with respect to an idol, nevertheless, is itself composed of matter and form and, therefore, bronze is not called prime matter, because it has matter. But that matter which is understood without any form and privation at all, but which is subjected to form and privation, is called prime matter, because there is no other matter before it. And this is also called 'hyle.'
>
> And because every definition and all cognition is through the form, therefore, prime matter as such (per se) cannot be known or defined except through a composite, so that it may be said that prime matter is what is related to all forms and privations in the way that bronze is related to an idol and unshaped. And this is called absolutely (simpliciter) prime. For something can be called prime matter with respect to some genus, as water is prime matter in the genus of liquefiables. It is not absolutely prime, however, because it is composed of matter and form. And so it has a matter that is prior.

> And we must also know that prime matter, and form too, are not generated or corrupted, because every generation is from something to something. But that from which there is generation is matter and that to which it is is form. If matter or form were generated, therefore, there would be matter of matter and form of form ad infinitum. And so, properly speaking, there is generation only of a composite.

And he adds:

> And we should also know that prime matter is said to be one in number in all things. But to be one in number is said in two ways: viz., what has one form determinate in number, e.g., Socrates, and prime matter is not said to be one in number in this way, since as such it has no form; something is also said to be one in number, which is without dispositions which make it differ in number. And it is in this way that prime matter is said to be one in number, because it is understood to be without all the dispositions which make for numerical difference.
>
> And we must also know that, granted that prime matter in its meaning does not have any form or privation, as in the meaning of bronze there is neither shaped nor unshaped, nevertheless, it is never devoid of form and of privation. But sometimes it has one form and sometimes another. But it can never exist as such (per se), because, since in its meaning it does not have any form, it does not have being in act, for there is no being in act except from form. But it exists in potency only. And, therefore, whatever is in act cannot be called prime matter.

The above lengthy quotation seems to be a superb placing of prime matter in the order of things. To sum up what was said, prime matter

1. is matter with no form or privation at all but is subjected to these;
2. is unknown as such, but known through a composite;
3. is not generated or corrupted, for this would be infinite regress;

4. is numerically one in all things;
5. is never devoid of form and privation, although it does not have these in its meaning.

In n. 1277 of the *In VII Metaph.*, St. Thomas, after repeating the example of Aristotle that in the case of a bronze statue bronze is as the matter and the figure or shape is as the form of the species, says:

> This example, indeed, is not to be taken strictly but in the similitude of proportion. For figure and other artificial forms are not substances but accidents. But because in this manner figure is related to bronze in artificial things, as substantial form is related to matter in natural things, he all the more deliberately uses this example to demonstrate the unknown by the known.

And in n. 1276, he says:

> He says, therefore, in the first place, that the subject, which is the first particular substance, is divided into three parts, viz., into matter, form, and what is composed of these. And this, indeed, is not a division of genus into species, but of some predicate that is analogously predicated according to priority and posteriority of those things that are contained under it. For the composite, as well as the matter and the form, are called a particular substance, but not in that order. And, therefore, he will investigate later which of these is primarily substance.

155. Here St. Thomas says that prime matter cannot be known in itself, but only analogously. And he specifies this, when he says "by proportion."

Prime matter is not readily or immediately disposed to exist but only mediately through form. We cannot know prime matter directly but only through analogy, i.e., by its resemblance to another object of our knowledge. The reason for this is that, since our knowledge is posterior to things, we know things from their acting upon our senses. But things do this to the extent that they are in act. Since the act of a sensible is its form, we can have no direct knowledge of prime matter; its very nature is to be uninformed but *for* form. Thus, matter is not knowable in and of itself but only analogously.

An analogy is a relationship midway between equivocity and univocity. And there are three kinds of analogy: attribution or proportion, inequality, and proportionality. An example of the first kind would be that of health as referred to an animal and grain. In this kind of analogy, the term predicated, i.e., health, is intrinsic to the animal but not to the grain. It is in the grain by extrinsic denomination only. The term health is neither completely the same nor completely other in its application to animal and to grain. Notionally, it is the same, but "existentially" it is not the same.

With regard to what some call the analogy of inequality, John of St. Thomas in his *Curs. Phil.*, Vol. II, qu. 13, art. 3 and 4 says that this is not an analogy strictly speaking.

The analogy of proportionality is the one which Cajetan says is the only true one. An example of such an analogy would be the predication of being of substance and of accident. The term being is intrinsic to both substance and to accident but as predicated of these, being is neither completely different nor completely the same notionally or "existentially." This latter term is our translation of "secundum esse" as used here in contrast to intentionally, our translation of "secundum intentionem."

Now, even though the analogy of attribution is sometimes called an analogy of *proportion*, it is obvious here in the *In Physic.*, where he says that prime matter is known "by proportion," that St. Thomas must mean an analogy of *proportionality* for, in the proportion that prime matter is to natural substances as bronze is to a statue or as wood is to a chair, the analogue, viz., a substrate, is *intrinsic* to all the analogates of which it is predicated. Prime matter is intrinsic to the existence of a tree and of a horse. Bronze is intrinsic to the existence of a bronze statue or monument. And the same is true of wood with reference to a chair or bed. In all cases, both the being and meaning of the analogue, as predicated, are neither completely the same nor completely different. The predication of prime matter of natural substances in the way that bronze is predicated of statues, or the proportion that prime matter is related to natural forms in the way that sensible matter is related to artificial forms, fulfills the pre-requisite for an analogy of proportionality as stated by St. Thomas in the *In I Sent.*, dist. 19, qu. 5, art. 2, ans. 1. This prerequisite is that the analogue be intrinsic to all the analogates and that the terms be analogous, i.e., neither purely univocal nor purely equivocal both in their being and in their being known.

In the above citation of *In I Sent.*, St. Thomas says:

> There are three kinds of analogy. One kind is *in the intentional order only and not in the order of existence*. An example of this would be when one intention is predicated of several objects according to priority and posteriority; this notion, however, is only *in* one of the objects, e.g., the notion of health is predicated of an animal, of urine and of a diet in different ways, i.e., according to priority and posteriority, but not according to a different act of being, because the being of health resides only *in* the animal.
>
> There is another kind of analogy *in the order of existence and not in the order of intention*. This happens when many things are reduced to a common denominator or note, which common note, however, does not have the same intelligible content in each object, e.g., all bodies are reduced to the common note of corporeity. As a result, the logician, who only considers intentions, claims that the word 'body' is predicated of all bodies univocally. Such being, however, is not essentially

> the same in corruptible and incorruptible bodies. Thus, for the metaphysician and the natural philosopher, who consider things under the aspect of their existence (secundum suum esse), neither the word 'body' nor any other predicate is attributed univocally of corruptible and incorruptible beings....
>
> The third kind of analogy would be that *according to both intention and existence*; and this occurs when a thing is neither reduced to a common intention nor to the same act of existence, as when being is predicated of a substance and of an accident. In such cases, the common nature must have some existence in each of the objects to which it is attributed, but this existence must be different insofar as the nature has the note of greater or lesser perfection in each object.
>
> And so, in this way I say that truth and goodness and all other like perfections are predicated analogously of God and of creatures. Hence, all these must be in God according to their very being, and they are in creatures insofar as they are in different creatures in a greater or lesser perfection. Thus, it follows that, since they cannot be with the same act of being in both cases, they are different truths.

And in the *De Ver.*, qu. 2, art. 11, St. Thomas expressly distinguished the two meanings of "by proportion." In one way, proportion may be said as many things have a proportion to one thing, as health is said of medicine and complexion. Both have an order and proportion to the health of a being, but medicine is a cause of health, while complexion is but a sign of health. In other words, health is only extrinsically predicated of complexion and of medicine. The second way in which we may speak of proportion, according to St. Thomas in the *De Ver.*, is as one thing has a proportion to another as health is said of medicine and of animal, inasmuch as medicine is the cause of health which is in the animal. The intention here is to consider medicine as ordered to the health of the animal and to consider health as formally intrinsic to animal. An analogy of proportion can, therefore, be called an analogy of intrinsic attribution, as in the last way, or an analogy of extrinsic attribution, as in the first way.

Since prime matter is intrinsic to the existence of all natural substances or natural forms, and if we are to know prime matter analogously, as St. Thomas says here in the *In Physic.*, then, we must know it by an analogy of intrinsic attribution, i.e., an analogy of proportionality, for it is in this latter kind of analogy that the analogue is formally intrinsic to all the analogates of which it is predicated. Actually, however, an analogy of intrinsic attribution simultaneously includes both proportion and proportionality. It includes the former because there is the relationship of cause to effect and it includes the latter because there is intrinsic denomination.

Lesson Fourteen

156. In less. 1, n. 338 of the *In III Metaph.*, St. Thomas shows us the importance of considering difficulties or doubts. He says that in seeking universal truth, which he intends to do in that work, we must first remove doubts. In nn. 339–342, he gives four reasons for considering doubts before determining the truth regarding what is in question. The *first* is that the investigation of truth is nothing other than the solution of previous doubts. *Secondly*, he repeats what he said in the beginning of the *In De Anima*, although it was stated there for a different reason, that we cannot determine the truth unless we first see the difficulty or difficulties involved. It is necessary to know where one is going before he asks the route to his destination.

Moreover, and this is the *third* reason, if we do not know the doubt beforehand, we will not know if and when we have arrived at its solution, i.e., if and when we have the truth that excludes the doubt. The *fourth* and last reason is that one is in a better position to judge the truth, if he has heard the contending arguments. One who hears all arguments is, according to Aristotle and St. Thomas, in a better position to form a judgment.

For these reasons, it was customary for these men, even though they did not take opposing opinions very seriously in influencing their own theories, to preface an investigation with a presentation of the doubts about the topic to be investigated. He announces this procedure again and again in the *In I Physic.* And here at the beginning of less. 14, we find him accomplishing one of the reasons why he starts out with the consideration of probables, for here he overtly says that through the truth that has now been determined, he is going to do away with the various doubts that the ancient philosophers had.

157. It might be well to indicate here that the other science is, of course, first philosophy or metaphysics, as he says later on in less. 15, n. 140.

158. For the ancients, there was no becoming from being because being already is, and also there was no becoming from non-being, because it has no subject.

159. The ancients were victims of this manner of speaking. We, however, say that if something cannot come to be in the strict sense, this does not mean that there can be no becoming incidentally.

160. This passage seems to be an attempt to bring forth and then to destroy a certain ambiguous way of speaking employed by the ancient philosophers.

There are two ways of speaking of a physician's doing or becoming something. Strictly speaking, we say that a physician may do something as a physician. Incidentally, however, he may do something not as a physician. Also, we may properly say that a physician may become something, e.g., a non-physician from his being a physician. He may also become something which is merely incidental to his being a doctor, e.g., he may become a good golfer.

According to the ancients, their line of reasoning would insist that there is only strict or proper becoming of either *something* or *being*, when that becoming is from non-being as such and that there is only incidental becoming from anything other than non-being as such. Aristotle and St. Thomas have parallelled "non-being as such" with "physician as physician," and they show later that there is strict becoming from something other than non-being as such.

The next paragraph of the text of St. Thomas brings out some of the consequences of this erroneous and ambiguous tenet with regard to strict becoming according to the views of the ancient philosophers.

161. The subject of the privation must be the same as the subject of the form. Otherwise, there would be a successive creation and not the becoming of one thing from another. The subject must remain.

In the *In XII Metaph.*, less. 3, n. 2443, we read:

> But that neither the ultimate form nor the ultimate matter are generated, he proves in this way: In every change there must be some subject of the change, which is matter, and something by which it is changed, which is the moving principle, and something into which it is changed, which is the species and the form. If the form itself and the matter were generated, e.g., if not only this whole, which is round bronze were generated, but also if the roundness itself and the bronze

> itself were generated, it would follow, therefore, that the form as well as the matter would have matter and form, and thus, we would proceed to infinity in matters and forms, which is impossible. And so, we must establish a starting point in generation, viz., that the ultimate form and the ultimate matter are not generated.

162. For the ancients, there was no becoming absolutely or in the full sense from being. Nor was there such becoming from any non-being whatsoever, but only from non-being as such, which was impossible.

If a dog comes to be from a horse, this is incidental becoming, since it is not coming to be from non-being as such but from another already existing animal. Even though the horse is non-dog, it is not non-animal or non-being absolutely.

Thus, for the ancients, there was no becoming in the full sense or unqualifiably, since, according to them, such becoming is from non-being as such. But every being, as fire or dog, comes from another being, as horse or air. And thus, there is the dilemma that there is no absolute becoming from being *or* from non-being.

It would be well to read the first five lessons of the first book of St. Thomas' commentary on the *De Gen. et Corr.* to see how the various ancient natural philosophers took becoming either as generation or as alteration. The next five lessons of that work show what Aristotle and St. Thomas consider to be generation in the full sense and in a restricted sense.

163. In less. 1, n. 1770 of *In IX Metaph.*, St. Thomas says:

> First Aristotle determines with regard to potency strictly speaking. Yet, this is not useful for the present purpose, for potency and act are, for the most part, said of what is in motion

> because motion is the act of a being in potency. But the main intention of this doctrine does not concern potency and act to the extent that they are in mobile things only, but to the extent that they follow upon being in general. Thus, there is potency and act in immobile things also, e.g., in intellectual beings.

We might add here by way of parenthesis that there is no such thing as local immobility in the case of separated substance, because there is no matter to be moved. Local mobility implies materiality which is a necessary requisite. There is no materiality in the separated substances, for they are out of the material world or material order.

The separated substances may be either substantial or accidentally immobile. God, the principal separated substance, is substantially and accidentally immobile. He is immobile both in His substance and in His operations. The angels have a substantial but not an accidental immobility. They are immobile with regard to their substance but not with regard to their operations. They have an immobility that is more than a local immobility.

St. Thomas continues in n. 1771:

> But when we shall have spoken about potency which is in mobile things, and about its corresponding act, we shall be able to explain further with regard to both potency and act insofar as they are in intelligible things, which pertain to separated substances.... And this is a logical order, since sensibles, which are in motion, are more known to us. And, therefore, through these we come to the cognition of the substances of immobile things.

With this introduction, St. Thomas endeavors to introduce the reader to the detailed analysis of potency and act which is considered throughout the *In IX Metaph*.

164. We read in less. 2, n. 2437 of the *In XII Metaph.*:

> He resolves a certain doubt.... Generation is change from non-being to being. But someone could wonder in what sense non-being is taken here, for we speak of non-being in three ways: in one way, that it in no way exists, and generation does not occur from such non-being, because nothing comes to be from nothing according to nature. In another way, non-being is called the privation itself, which is considered to be in some subject. And from non-being in this sense there occurs a certain kind of generation. But it occurs incidentally, i.e., to the extent that generation comes to be from a subject in which there happens to be privation. In the third way, non-being is called the matter itself, which, as such, is not being in act, but being in potency. And from such non-being generation comes to be in the full sense.

Lesson Fifteen

165. Becoming in the full sense is substantial becoming. Any other kind of becoming is becoming in a limited sense, which is accidental becoming. From prime matter and substantial form a substance comes to be. This is becoming "per se." From the substance that now has existence, other things may come to be. The substance may become white or musical. This would involve becoming in a limited sense or accidental becoming and not becoming "per se." Substantial becoming has as its substrate prime matter, which is in potency to substantial being, while accidental becoming has as its substrate an existing subject that is definite, and not prime matter which is unformed and indefinite. In substantial becoming the form gives being to something indefinite; in fact, it gives a thing its very being. In accidental becoming the accidental form further modifies something that already exists.

For the ancients there was no becoming in the full sense except from non-being as such. For St. Thomas and Aristotle there is becoming in the full sense or the becoming of substantial being when this becoming is from matter under one form to matter under another form. The former has the privation of the form that it is going to have. Privation is non-being, but the becoming is not from the privation, but from the prime matter which has the privation.

166. For the ancients, since privation, which is non-being, was attributed to matter and since they did not distinguish between unity in subject and in meaning, matter and privation were identified. Since privation is non-being as such, they believed that becoming in the full sense would have to be from this non-being and not from matter to which the privation was attributed.

167. Cf. Appendix note 164.

168. Since Bk. III of this work is concerned with the definition of motion, St Thomas starts out by prefacing some requisites for the investigation of the definition of motion before he defines it. In less. 1, n. 279, he says, referring to Aristotle:

> First, he premises certain divisions, since the best way of forming definitions is through divisions, as the Philosopher makes clear in the *Posterior Analytics* II and in the *Metaphysics* VII. Secondly, he shows that motion comes under the foregoing divisions.

And he continues in n. 280:

> With regard to the first, he states three divisions. The first of these is that being is divided by potency and act. And,

indeed, this division does not distinguish the genera of being, for potency and act are found in any genus whatsoever.

That this division of being into potency and act must precede the consideration of the definition of motion is made clear in n. 285 of less. 2 of the third book:

> And it is, therefore, entirely impossible to define motion by the prior and more known otherwise than as the Philosopher defines it here. For it has been said that every genus is divided by potency and act. But potency and act, since they are of the first differences of being, are naturally prior to motion, and the Philosopher uses these for defining motion.
>
> We must note, therefore, that there is something which is in act only, and something which is in potency only, and something which is intermediate between potency and act. What is in potency only is, therefore, not yet in motion, while what is already in perfect act is not being moved, but is already moved. And so, that is moved which is intermediate between pure potency and act. And this, indeed, is partly in potency and partly in act, as is obvious in the case of alteration. For, when water is only potentially hot, it is not yet moved, but when it is already heated, the motion of heating is terminated. But, when something is warming but in the state of imperfect heat, it is moved towards the state of being hot, for what becomes hot gradually participates more and more in heat. The imperfect act of heating existing in the thing to be heated is motion, not, indeed, as it is in act only, but as it, now existing in act, has a relation to a further act. Because if the relation to a further act were taken away, the act itself, however imperfect it might be, would be the terminus of motion and not motion, as happens when something becomes only partially hot. But the order to further act belongs to that which is in potency to it.
>
> And, likewise, if imperfect act were considered in relation to a further act only, accordingly it would have potency but not motion, but a principle of motion, for becoming hot can begin from a cold state as well as from a lukewarm one.

> Imperfect act, therefore, has the meaning of motion whether compared as potency to further act or to something more imperfect as act.
>
> Hence, it is neither the potency of something existing in potency, nor is it the act of something existing in act but it is the act of something existing in potency, so that through the word act, its order to anterior potency is designated, and through what is said to be in potency to the object that exists, its order to further act is designated.
>
> Hence, the Philosopher very fittingly defines motion by saying that it is the entelechy, i.e., the act of something existing in potency to the extent that it is of this nature.

All of Bk. III is somehow involved in an application of the definition of th nature of potency and act. In less. 3, after showing that motion is not otherness inequality or non-being, he says that these views were erroneously maintaine because of the indefinite and indeterminate nature of motion. Indeed, what motio is is most difficult to grasp.

In n. 296 St. Thomas, referring to Aristotle, says:

> He gives the reason why motion is placed in the order of the indeterminate.
>
> And he says that the reason for this is that it can be placed neither under potency nor under act. If it were placed under potency, whatever was in potency, e.g., to quantity, would be moved quantitatively. And if it were placed under act, whatever was quantity in act would be moved quantitatively. And it is true, indeed, that motion is an act, but it is an imperfect act intermediate between potency and act. And that it is an imperfect act is clear from the fact that that of which there is act is being in potency.... It is difficult, therefore, to understand what motion is, for it seems at first glance that it is either absolute act or absolute potency, or that it is contained under privation, as the ancient philosophers stated that it was contained under non-being and under equality. But none of these is possible.

Motion is neither absolute act nor absolute potency. Nor is it non-being or equality. Motion actuates potency without making it cease to be potency to the very same act. St. Thomas says in the last part of the above citation of the *In Physic.*:

> And so, for defining motion, there remains only the way as stated above, i.e., motion is an act such as we have said. It is the act of something existing in potency. But it is difficult to consider such act on account of the mixture of act and potency. That such act exists, however, is not impossible but actually occurs.

169. Unity of subject means that a subject is one in itself, although this subject may be looked upon as diverse in relation to the diversity of forms to which it may be related. Numerical unity or unity of subject is a unity of what is in itself undifferentiated. Unity in meaning, on the other hand, is a way of signifying the subject which includes all the other ways of signifying it. The meaning is one and only one way of signifying that which *may* be signified in other ways. A bronze statue is one numerically or in subject, but it is not one in meaning. We may speak of it as hard, shaped, in potency to be reshaped, cold, in potency to be warm, facing east, in potency to face north and so on. These are different significations or different meanings referred to what is one in itself or in subject, viz., the statue.

The potency of prime matter is one in subject. It is one potency, but it is in potency to a multiplicity of things. We can think of it as related to different forms. Thus, it has different meanings, if we refer to the various forms which will activate potency, e.g., the form man, dog or turnip.

170. It does not come as any great surprise to learn that the capacity to be cured and the capacity to be sick differ in intention or meaning. The references made to *In III Physics.*, is to less. 2, n. 290. In this lesson, St. Thomas analyzes the parts of the definition of motion, and in n. 290, he gives as example taken from contraries to show that "insofar as it is in potency" is necessary to the definition of motion:

> It is obvious that some same subject is in potency to contraries, as humor or blood is the same subject potentially disposed to health or to sickness. But it is obvious that to be in potency to health and to be in potency to sickness are different (and I mean this according to the order of objects). Otherwise, if the capacity to work and the capacity to be cured were the same thing, it would follow that to work and to be cured would be the same. The capacity to work and the capacity to be cured differ, therefore, in meaning, but they are one and the same in subject.
>
> It is evident, therefore, that a subject does not have the same meaning insofar as it is a certain being and insofar as it is in potency to something else. Otherwise, a potency to contraries would be one in meaning. And in this way, too, color and the visibles are not the same in meaning. And it was necessary, therefore, to say that motion is the act of something in potency *insofar as it is in potency*, lest one might understand it to be the act of that which is in potency, inasmuch as it is a certain subject.

Motion is not a subject, but it is the act of a subject. And motion is the act of a subject, not as subject, but as a subject in potency, i.e., not yet completely in act.

171. Here being one in potency is the same as being one in meaning. He is opposing unity in subject, i.e., numerical unity, with unity in meaning or a unity

hat is not an actual one. This latter kind of unity is only potentially one, since it ncludes, besides the essence of a thing, all the accidents that a being may have ıow and that it could have. And so the unity of meaning or intention, which would nclude the essential and accidental parts of a thing, would have but a potential and ıot an actual unity. There would be many parts forming a potential unity only.

172. Privation, which is a negative principle, does not enter into the constitution ɔf a substance of which it is a principle. This is what is meant by a negative ɔrinciple. The absence of a pilot may be the "cause" of a ship's foundering, yet, :his cause is not a positive influence upon the event which takes place.

173. Cf. Appendix notes 86, 127.

174. It may be clear that Plato and Aristotle posited principles in a way completely different, but here it does not seem clear that the principles themselves are entirely or completely different.

Plato: great-and-small → form
Aristotle: matter (privation) → form

Plato says that the principle that is the material element is one both in subject and in meaning, whereas Aristotle states two principles on the side of matter. One of these is a substantial principle, viz., prime matter, and the other is an accidental principle, which does not enter into the constitution of the thing made.

Now, although Plato's great-and-small was considered as one in subject and in meaning, still there was a duality. Plato did not state privation as a principle, yet his great-and-small is an indefinite, unlimited principle, capable of infinite limitations by form. In the way that Plato stated this principle, he did not take into account the negative principle of privation, yet, he seemed to include the general meaning of Aristotle's prime matter *and* privation.

From the viewpoint of the form as a principle for both Aristotle and Plato the prime matter or great-and-small were pliable and uninformed until limited by a form. Form, in the case of the theory of both Plato and Aristotle, was that by which the indefinite became definite. For Plato, however, limitation by form was accomplished through participation in the form, a separated form. Aristotle's idea of form was that it was the act through which prime matter was activated to be a certain being that has such act or such form. This form, however, was not a separated form or species, as it was for Plato.

Fundamentally, it seems that the general tenor and intention of the principles as posited by these men are similar in some respects.

175. Form "is divine, indeed, because every form is a certain participation in a similitude of divine being, which is pure act, since a thing is in act to the extent that it has form." We feel that in these few lines there is given not only the nucleus and perhaps the culmination of the philosophy of St. Thomas but also the refreshing nobility of the philosophy of nature and the beings which it studies. These lines, or in fact, this one sentence of St. Thomas seems to us to be the most potent, sublime and challenging sentence in all of this first book. The meaning in this statement allows natural beings to take their place in a great order and ordering. They are not crude entities devoid of import. They take on a kind of transcendence. "Every form," St. Thomas says, participates in a similitude of the divine. The string bean, the turtle and man have a greater meaning and intelligibility by reason of their participation, through their forms, in the similitude of the

divine being. Matter should not be scorned. It is the medium that receives the forms through which the entities thus composed assume an inalienable sublimation.

176. In the Aristotelian teleology of nature, the higher a being is in the hierarchy of beings, the more perfect will be its activity, for the kind of operation a thing has depends upon the kind of being that it is. And a being is what it is by reason of its form.

According to Aristotle, the order in the universe demands that each being seek its good in an hierarchical system. The lower seeks the higher, for the ultimate purpose of the universe is the unmoved mover. Form is the good of a thing and is, therefore, desirable. A being, then, seeks its good as its perfection. A methodology of teleology permeates the whole Aristotelian-Thomistic doctrine.

177. Matter implies an aptitude to and a capacity for form, and since the form is its good, the matter, then, is not evil. In the *In I Peri Hermeneias*, St. Thomas says that, although privation may pertain to evil, matter does not. In less. 11, n. 151, par. 2, he says:

> To the extent that matter is taken as worse, it is true, according to the opinion of Plato, who did not distinguish privation from matter, but it is not true, in the opinion of Aristotle, who says in the first book of the *Physics* that evil and base and other such things pertaining to defect are not said of matter, except in an incidental way....

178. It is vital here to make a distinction between two ways of considering prime matter. The indefinite pure potency or passive capacity to receive form has of itself no dynamic tendency to seek form. It is mere ordering to form. We can

only metaphorically say that such matter "desires" form. The second manner of considering matter is that of prime matter as the ultimate material constituent or substrate in an already existing entity. This "informed" matter, which is sometimes called second matter, may be considered apart from its form. According to Augustin Mansion, p. 241, *op. cit.*, this matter "is no longer pure indetermination, but it is in potency with regard to a further determination." Again, it does not of itself actively "desire" or "seek" after further determinations or the acquisition of further being, but it does this only indirectly through form. Matter of itself, whether it be prime matter in general or the kind of prime matter that is called second matter, is only a passive but an indispensable principle of receiving forms. But second matter or matter that is "informed" has a dynamic tendency through the form to further perfection and the acquisition of further being. In other words, matter through "information" acquires a tendency towards further "information." Such matter has the capacity for all the forms to which it is in potency and always desires a form that it does not have until this tending achieves the ultimate end of the whole process of generation to which it is ordered. And this ultimate end is the human soul. The tending of matter to form and further information is always that of matter that is informed, and certainly not that of prime matter that is sheer "uninformedness." And for those beings that are informed by a human soul, there is again the tending to perfect that soul through knowledge. The human soul is the highest kind of being apart from angelic and divine being. But this being, informed by this soul, strives for perfection through the perfection of its highest faculty, the intellect, which becomes in an intentional way what it knows. And it has the capacity to become intentionally all being. According to Rousselot, *op. cit.*, the intellect is an acquisitive faculty and is the *raison d'être* of all forms of appetite. In the terrestrial order, this would make knowledge or the tending to knowledge the culmination of the tendency of informed matter to further form. From here we could easily proceed further, if we were to take the order of the divine into more specific consideration, and make contemplation this culmination, but this might transcend the order of natural philosophy.

That there is a constant tendency in nature to the ultimate end of generation, viz., the human soul, St. Thomas makes clear in the *Cont. Gent.*, Bk. III, chap. 22:

> What are moved or acted upon only without moving or doing anything of themselves tend towards the divine similitude to the extent that they are perfected in themselves. But what act and move, as such, tend towards the divine similitude in that they are the causes of others. And what move by being moved tend to the divine similitude in both these ways....
>
> Prime matter tends to its perfection by acquiring in act a form that it previously had in potency, although it may cease to have the other form that it previously possessed in act. For it is in this way that matter successively receives all the forms to which it is in potency, in order that all of it may be successively reduced to act, which is something that could not be done all at once.

And he continues in this same chapter to show how it is that prime matter is in constant tendency to form up to the end of the process of generation to which it is ordered:

> Whatever is moved, to the extent that it is moved, tends to the divine likeness so that it may be perfected in itself. But a thing is perfect to the extent that it is in act. The intention of everything that exists in potency must be to tend to act through movement. The more an act is posterior and perfect, therefore, the more principally is the appetite of matter directed towards it. Hence, regarding the last and most perfect act that matter can attain, the appetite of matter by which it seeks form must

> tend as to the ultimate end of generation. But in the acts of the forms there are various gradations. For prime matter is first in potency to the form of an element. When it has the form of an element, it is in potency to the form of a mixed body, because elements are the matter of a mixed body. Considered as having the form of a mixed body, it is in potency to a vegetative soul, for this is the soul that is the act of such a body. Likewise, the vegetative soul is in potency to a sensitive soul, and the sensitive soul to the intellectual soul. The process of generation makes this clear, for first in generation there is the living foetus possessing the kind of life proper to a plant, later that of animal life, and finally the life of a man. No later or more noble form is found in generable and corruptible things after the last form, i.e., the soul of a man. The ultimate end of all generation is, therefore, the human soul and matter tends to this as its ultimate form. Elements, therefore, are for the sake of mixed bodies, but these latter are for the sake of living bodies. In these latter, plants are for the sake of animals; but animals are for the sake of man. Man, therefore, is the end of all generation.

There is, then, a hierarchy of "tendings" of matter to form. Because of its indetermination, matter is in potency to all the forms that it can have, but there is an order of receiving these forms. It is always in potency to a higher form, until it has reached the ultimate form that it can have, viz., the human soul.

179. These philosophers would have privation desire form, which would mean that a contrary desired its contrary and, consequently, its self-destruction. The truth, however, is that informed matter is the capacity for a form which implies the absence of the form that it has; and through form it tends towards other forms. Privation does not tend towards form, for privation is non-being purely and simply and, as such, cannot tend towards what is contrary to it, viz., form.

A thing does not desire its contrary but rather it desires what is like to it in nature. In the *De Ver.*, qu. 22, art. 1, ans. 3, St. Thomas says:

> Whatever desires anything does so to the extent that it has some likeness to it.... And so, the likeness must be of its nature. But this kind of likeness may be spoken of in a two-fold way: in one way, as the form of one thing is in perfect act in another. And from the fact that something is thus likened to the end, it does not tend to the end but rests in it. In another way, from the fact that the form of one is in the other in an incomplete manner, i.e., in potency, and thus, insofar as something potentially has the form of the end and of the good in it, it tends to the good or the end and desires it.
>
> And in this way, matter is said to desire form to the extent that form is potentially in it. And, therefore, the more that potency is more perfect and closer to act, the more it causes a more vehement inclination. From this, it so happens that all natural motion is intensified near the end when what tends to the end is now more like the end.

180. Just as matter does not have capacity for the form that it now possesses, since it already has it, it does not desire a form contrary to the form that it has, but one contrary to the privation that it has.

If matter and privation were the same, privation would seek form and thus would seek its own destruction. Privation is a lack or denial of something. Form is the something that is lacking. This lack or negation in a determinate subject is destroyed by a positive reality. Privation, then, is destroyed by substantial form, which is something positive. Although one form may destroy a given privation, it is still true that a subject may have other privations or privations of *other* forms which may, in turn, be destroyed by the acquisition of these other forms.

181. It is precisely because prime matter lacks form that we have a difficult time trying to know anything about it. There is evidence of the struggle to know something through "unformedness" in the *De Pot. Dei*, qu. 4, art. 1, ans. 2, where St. Thomas says:

> But according to Augustine...through the names of earth and of water, which are mentioned before the formation of light, are not understood the elements formed from these forms but unformed matter itself, which lacks every species. But for that reason the unformedness of matter is expressed through these two rather than through other elements, since they are closer to unformedness in that they have more matter and less form. And these, which are even better known to us and more obvious to us, clarify the matter of the others. For that reason, however, he wanted it to be signified not through one only but through two, lest if he had posited the other only, he would really have believed that this was unformed matter.

The quotation above is clear in that it makes the idea "unformedness" of matter to exceed even the seeming "unformedness" of either earth or of water before creation.

182. According to Aristotle and St. Thomas, wherever there is orientation towards something, it follows upon knowledge either in the subject or in another which orients it.

183. There is basis here for the dynamism of change according to St. Thomas. Mutability in an already existing being is not merely *potency to* change, but it is a *tendency to* change, for St. Thomas says that even though a thing may have *one form in act*, it still has potency to *and appetite for* other forms. This passage

would disprove those who disclaim any dynamism in the potency-act theory of St. Thomas.

184. The reference here to Plato's use of this metaphorical language is to the *Timaeus*, 50–51. The citation which follows shows that Plato's understanding of a receiving principle or substrate is closer to "matrix" than to matter as we think of it.

> For the present we have only to conceive of three natures: first, that which is in process of generation; secondly, that in which the generation takes place; and thirdly, that of which the thing generated is a resemblance. And we may liken the receiving principle to a mother, and the source or spring to a father, and the intermediate nature to a child; and may remark further, that if the model is to take every variety of form, then the matter in which the model is fashioned will not be duly prepared, unless it is formless, and free from the impress of any of those shapes which it is hereafter to receive from without. For if the matter were like any of the supervening forms, then whenever any opposite or entirely different nature was stamped upon its surface, it would take the impression badly, because it would intrude its own shape. Wherefore, that which is to receive all forms should have no form; as in making perfumes they first contrive that the liquid substance which is to receive the scent shall be as inodorous as possible; or as those who wish to impress figures on soft substances do not allow any previous impression to remain, but begin by making the surface as even and smooth as possible.
>
> In the same way that which is to receive perpetually and through its whole extent the resemblances of all eternal beings ought to be devoid of any particular form. Wherefore, the mother and receptacle of all created and visible and in any way sensible things, is not to be termed earth, or air, or fire, or water, or any of their compounds or any of the elements from which these are derived, but it is an invisible and formless being

> which receives all things and in some mysterious way partakes of the intelligible, and is most incomprehensible.
>
> In saying this we shall not be far wrong; as far, however, as we can attain to a knowledge of her from the previous considerations, we may truly say that fire is that part of her nature which from time to time is inflamed, and water that which is moistened, and that the mother substance becomes earth and air, in so far as she receives the impressions of them.

185. It is strange that St. Thomas uses the word "contra" here, for there seems to be no reason for Aristotle's making fun of Plato for using the example of a mother in his attempt to make clear the nature of matter. Plato has approached the meaning of matter in a more philosophical manner elsewhere.

186. One might ask why we cannot *directly* refer natural forms to separated substances, in the manner that the Platonists did. The species of natural forms include both a natural form and matter, and thus belong directly to natural philosophy, while the species of separated substances are separated forms without matter, and are not within the scope of this science. In the *De Spiritualibus Creaturis*, qu. 1, art. 1, St. Thomas explains the hierarchy of natural and separated substances.

> In composite things, there are two kinds of act and two kinds of potency. In the first place, matter is as potency in reference to form, which is its act. Secondly, if the nature is made up of matter and form, the matter is as potency with reference to existence itself to the extent that it, i.e., the matter, is able to receive this. And so, when the foundation of matter is taken away, if there remains any form of a determinate nature which is self-subsistent but not in matter, it will still be related to its own existence *as potency to act*. But I do not say *as that*

> *potency which is separable from its act*, but *as a potency is always accompanied by its act*. And thus, the nature of a spiritual substance, which is not made up of matter and form, is a potency with reference to its own existence. Thus, in a spiritual substance, there is a composition of potency and act, and, consequently, of matter and form, i.e., if every potency is called matter and every act is called form. But this is not said properly according to the common use of these words.

187. Throughout the first book of the *In Physic.*, St. Thomas has been concerned with the *principles of natural beings*. These principles were established and they are matter, form and privation. Now, in the second book, it is the *principles of the kind of knowledge* involved in natural philosophy that are studied.

BIBLIOGRAPHY

Books

Allen, T. *The Quest*. New York: Chilton Press, 1965.

Aristarchus. *Aristarchus of Samos*. Translated by Sir Thomas Heath. Oxford: Clarendon Press, 1913.

Aristotle. *Categoriae, De Caelo, Peri Hermeneias, Priora Analytica, Topica*. Contained in *The Basic Works of Aristotle* by Richard McKeon. New York: Random House, 1941.

________. *Physica*. Trans. by Philip H. Wicksteed and Francis M. Cornford. *Aristotle: The Physics*. New York: G. P. Putnam's Sons, 1929.

________. *Physics*. Translated by Richard Hope. Lincoln, Neb.: University of Nebraska, 1961.

Armstrong, Neil. *First on the Moon*. Epilogue by Arthur Clarke. Boston: Little Brown and Company, 1970.

Asimov, Issac. *Earth is Room Energy*. New York: Doubleday and Company, 1957.

________. *The Gods Themselves*. Greenwich, Conn.: Fawcett Publishing Company, 1972.

________. *The Solar System and Back*. New York: Doubleday and Company, 1970.

________. *The Stars Like Dust*. London: Sidgwick and Jackson, 1969.

Asimov, Issac. *Today and Tomorrow*. New York: Doubleday and Company, 1973.

________. *The Universe: From Flat Earth to Quasar*. New York: Discus Books, 1968.

Bachelard, Gaston. "La Connaissance Approchée." *Gaston Bachelard*. Paris: J. Vrin, 1927.

________. *...L'eau et les rives*, essai sur l'imagination de la matière. Paris: J. Corti, 1942.

________. "On Poetic Imagination and Reverie." *Selections from the Works of Gaston Bachelard*. Translated and introduced by Colette Gauden. Indianapolis: Bobbs-Merrill, 1971.

________. *The Philosophy of No*. Translated by G. C. Waterston. New York: Orion Press, 1968.

________. *The Poetics of Reverie*. Translated by Daniel Russell. New York: Orion Press, 1969.

________. *The Poetics of Space*. Translated by Maria Jolas. Foreword by Etienne Gilson. New York: Orion Press, 1964.

________. *The Psychoanalysis of the Fire*. Translated by Alan C. M. Ross. Boston: Beacon Press, 1964.

________. *The Right to Dream*. Translated by J. A. Underwood. New York: Grossman, 1921.

Beaufret, Jean. *Le Poème de Parménide*. Paris: Presses univérsitaires de Paris, 1955.

Bentham, Jonathan. *Science and Technology in Art Today*. New York: Praeger, 1972.

Bergson, Henri. *Creative Evolution*. Translated by Arthur Mitchell. London: Macmillan Company, 1954.

Berman, Louis. *Exploring the Cosmos*. Boston: Little Brown and Company, 1973.

Bernard, Claude. *Introduction à l'étude de la médecine expérimentale*. Paris: Delgrave, 1938.

Berrill, N. J. *Worlds Without End*. New York: The Macmillan Company, 1964.

Bertalanffy, Ludwig Von. *Robots, Men and Minds*. New York: Braziller, 1967.

Bertholon, Pierre. *De L'Electricité Des Végétaux*. Paris, 1783, 5-Microprints, Readex Microprints, New York, 1968.

Bester, Alfred. *The Stars My Destination*. New York: New American Library, 1957.

Bova, Benjamin. *Planet, Life and L. G. M.* (Little Green Men). Reading, Mass.: Addison-Wesley, 1970.

Bradbury, Ray. *I Sing the Body Electric*. New York: Doubleday and Company, 1969.

________. *Mars and the Mind of Man*. New York: Harper and Row, 1973.

________. *Martian Chronicles*. New York: Doubleday and Company, 1958.

________. *S is for Space*. New York: Doubleday and Company, 1966.

Bruno, Giordano. *The Cosmology of Giordano Bruno*. Translated by K. E. W. Maddison. Ithaca, N.Y.: Cornell University Press, 1973.

________. *Heroic Enthusiasts*. St. Clair Shores, Mich.: Scholarly Press, 1887.

________. *The Infinite Worlds of Giordano Bruno*. Springfield, Ill.: Charles C. Thomas, 1970.

Burnet, John. *Early Greek Philosophy*. London: Adam and Charles Black, 1948.

________. *Greek Philosophy from Thales to Plato*. London: Macmillan, 1914.

Burroughs, Edgar. *At the Earth's Core*. New York: Ace Publishing Corporation, 1973.

Cade, C. M. *Other Worlds than Ours*. New York: Taplinger, 1967.

Cajetan, Thomas de Vio. *De Analogia Nominum*. Literally translated and annotated by Edward Bushinski in *The Analogy of Names*. Pittsburgh: Dusquesne University Press, 1953.

________. *In Summam Theologicam*. Contained in *Opera Omnia Sancti Thomae*. Rome: Leonine, 1888.

Cameron, A. G. W. *Interstellar Communications*. New York: W. A. Benjamin, Inc., 1963.

Carden, Martin. *Destination Mars*. New York: Doubleday and Company, 1972.

Clarke, Arthur. *Childhood's End*. New York: Harcourt, Brace and Jovanovich, 1953.

________. *The Coming of the Space Age*. London: Gollanez, 1967.

________. *Profiles of the Future*. New York: Harper and Row, 1962.

________. *Rendezvous with Rama*. New York: Harcourt, Brace and Jovanovich, 1973.

Cole, D. W. and Scarfo, R. G. *Beyond Tomorrow*. Amherst, Mass.: University of Massachusetts Press, 1965.

Comte, August. *Introduction to Positive Philosophy*. Indianapolis, Ind.: Bobbs-Merrill Company, 1970.

Copleston, Frederick, S.J. *History of Philosophy*, Vol. I: *Greece and Rome*. Westminster, Md.: The Newman Bookshop, 1946.

Cornford, Francis M. *Plato and Parmenides*. New York: Harcourt Brace and Company, 1939.

Crichton, Michael. *The Andromeda Strain*. First edition. New York: Alfred A. Knopf, Inc., 1969.

Darwin, Charles. *The Next Million Years*. Westport, Conn.: Greenwood Press, 1973.

De Bergerac, Cyrano. *Voyage to the Moon and Sun*. Translated by Richard Addington. New York: Orion Press, 1962.

De Broglie, Louis. *Matière et lumière*. Paris: A. Michel, 1937.

Dole, S. H. and Asimov, I. *Planets for Man*. New York: Random House, 1964.

Dole, Stephen. *Habitable Planets for Men*. New York: American Elsevier, 1970.

Dougherty, Kenneth. *Cosmology: An Introduction to the Philosophy of Nature*. Peekskill, N.Y.: Graymoor Press, 1952.

Doyle, Conan. *The Disintegration Machine*. London: Murray, 1952.

Drake, F. D. *Intelligent Life in Space*. New York: The Macmillan Company, 1962.

Drake, Stillman. *Discoveries and Opinions of Galileo*. New York: Doubleday and Company, 1957.

Duhem, Pierre. *La théorie physique: son objet et sa structure*. Paris: Rivière, 1914.

Eddington, Arthur. *Nature of the Physical World*. New York: Macmillan, 1927.

Eddington, Arthur. *The Philosophy of Physical Science*. Ann Arbor: The University of Michigan Press, 1958.

Ehrike, Krafft A. *Space Flight*. Princeton, N.J.: Van Nostrand, 1960.

Einstein, Albert and Infeld, Leopold. *The Evolution of Physics: The Growth of Ideas from Early Concepts of Relativity and Quanta*. New York: Simon and Schuster, 1938.

Eiseley, Loren. *Firmament of Time*. New York: Atheneum Press, 1970.

________. *The Invisible Pyramid*. New York: Charles Scribner's Sons, 1970.

________. *The Unexpected Universe*. New York: Harcourt, Brace, Jovanovich, 1964.

Eliade, Mircea. *Myth and Reality*. Translated by Willard R. Trask. New York: Harper and Row, 1963.

________. *The Myth of Eternal Return*. Translated by Willard R. Trask. New York: Harper and Row, 1959.

Fallacci, Oriana. *If the Sun Dies*. Translated by Pamela Swinglehurst. First edition. New York: Atheneum Press, 1966.

Firsoff, V. A. *Life Beyond the Earth*. New York: Basic Books, 1963.

Frank, Philipp. *Foundation of Physics*. Chicago: University of Chicago Press, 1946.

Freeman, Kathleen. *Ancilla to the Pre-Socratic Philosophers*. Oxford: Basil Blackwell, 1956.

________. *Pre-Socratic Philosophers*. Oxford: Basil Blackwell, 1949.

Galileo, G. *Dialogue Concerning the Two Chief World Systems*. Translated with notes by S. Drake. Berkeley: University of California Press, 1962.

Gardeil, H. D., O.P. *Introduction to the Philosophy of St. Thomas*. Vol. II: *Cosmology*. Translated by John A. Otto. St. Louis: B. Herder Book Company, 1948.

Gatland, Kenneth and Dempster, Derek. *The Inhabited Universe*. New York: Fawcett Publications, Inc., 1959.

Geiger, O., O.P. *La Participation dans la philosophie de St. Thomas d'Aquin*. Paris: J. Vrin, 1953.

Goddard, R. H. *Papers of Robert H. Goddard*. New York: McGraw-Hill Book Company, 1969.

Gray, J. Glenn. *Hegel and Greek Thoughts*. New York: Harper and Row, 1968.

Haldane, J. B. S. *Daedaluls; or Science and the Future*. New York: E. P. Dutton, 1924.

Heinlein, Robert A. *The Past Through Tomorrow*. New York: G. P. Putnam's Sons, 1967.

Hetzler, Florence M. *Introduction to Natural Philosophy*. Cambridge, Mass.: Harvard University Library, 1959. Microfilm.

Hipparchus, B. *The Geographical Fragments of Hipparchus*, D. R. Dicks, editor. London: University of London, Athlone Press, 1960.

Hoenen, Peter, S.J. *The Philosophical Nature of Physical Bodies*. Parts One and Two from Book Four of the *Cosmologia*. Translated by David Hassel, S.J. West Baden Springs, Ind.: West Baden College, 1955.

Hynek, Allen. *Astronomy One*. Menlo Park, Ca.: W. A. Benjamin, Inc., 1972.

________. *The UFO Experience*. Chicago: Henry Regnery Company, 1972.

Jaffa, Harry V. *Thomism and Aristotelianism*. Chicago: University of Chicago Press, 1952.

Jastrow, Robert. *Red Giants and White Dwarfs*. New York: New American Library, 1967.

John of St. Thomas. *Cursus Philosophicus Thomisticus*. Reiser edition. Turin: Marietti, 1930–1933.

________. *Cursus Theologicus*. Solesmes edition. Paris: Desclée, 1931.

Jones, Harold Spencer. *Life on Other Worlds*. New York: The Macmillan Company, 1940.

Knowles, David. *The Historical Content of the Works of St. Thomas Aquinas*. A paper read to the Aquinas Society of London in 1956. London: Blackfriars, 1958.

Koestler, Arthur. *The Sleepwalkers*. New York: The Macmillan Publishing Company, Inc., 1968.

LaPlace, Pierre S. *Celestial Mechanics*. Bronx, N.Y.: Chelsea Publishing Company, 1969.

Le Maître, George E. *Beaumarchais*. New York: Alfred A. Knopf, Inc., 1949.

Leonard, William Ellery. *The Fragments of Empedocles*. Chicago: The Open Court Publishing Company, 1908.

Levy, Lillian. *Space: Its Impact on Man and Society*. New York: W. W. Morton and Company, 1965.

Lovell, Bernard. *Our Present View of the Universe*. Manchester, England: Manchester University Press, 1967.

MacGowan, R. A. and Ordway, F. I. *Intelligence in the Universe*. Englewood Cliffs, N.J.: Prentice-Hall, Inc., 1966.

Mansion, Augustin. *Introduction à la physique aristotélicienne*. Paris: J. Vrin, 1946.

Maritain, Jacques. *The Degrees of Knowledge*. New York: Scribner's, 1938.

________. *The Philosophy of Nature*. New York: Philosophical Library, 1951.

McWilliams, James, S.J. *Physics and Philosophy*. Washington: Catholic University of America, 1936.

Mesmer, F. A. *Maxims on Animal Magnetism*. Translated by Jerome Eden. Mount Vernon, N.Y.: Eden Press, 1958.

Mitchell, Edward. *How He Will Reach the Stars*. New York: The Macmillan Publishing Company, Inc., 1969.

Moffat, S. and Schneour, R. A. *Life Beyond the Earth*. New York: Four Winds Press, 1966.

Mumford, Lewis. *The Myth of the Machine*. New York: Harcourt, Brace, Jovanovich, Inc., 1967.

________. *The Transformation of Man*. New York: Harper and Row, 1956.

Niebuhr, Reinhold. *Moral Man and Immoral Society*. New York: Charles Scribner's Sons, Inc., 1960.

Ostrander, Sheila and Schroeder, Lynn. *Psychic Discoveries Behind the Iron Curtain*. Englewood Cliffs, N.J.: Prentice-Hall, 1970.

Ovender, Michael W. *Life in the Universe*. Garden City, N.Y.: Doubleday and Company, 1962.

Parc, Ambroise. *Apologies and Treatise of Ambroise Parc*. New York: Dover Publications, Inc., 1968.

Piers, Anthony. *Macroscope*. New York: Avon Books, 1969.

Phillips, R. P. *Modern Thomistic Philosophy*. Vol. I: *The Philosophy of Nature*. Westminster, Md.: The Newman Press, 1950.

Plato. The following dialogues: *Laws, Parmenides, Phaedo, Philebus, Republic, Sophist, Statesman, Timaeus*. The references to these works have been taken from the translations by Benjamin Jowett. *The Dialogues of Plato*. Vols. I and II. New York: Random House, 1920.

Poincaré, Henri. *Science and Hypothesis*. New York: Dover Publications, Inc., 1905.

Raven, J. E. *Pythagoreans and Eleatics*. Cambridge: The University Press, 1948.

Reiff, Philip. *The Triumph of the Therapeutic*. New York: Harper Torchbook, 1961.

Reiser, Oliver. *Cosmic Humanism*. Cambridge, Mass.: Schenkman Publishing Company, 1966.

Robin, Léon. *La théorie platonicienne des idées et des nombres d'après Aristote*. Paris: F. Alcan et Guillaumin, 1908.

Robinson, Richard. *Plato's Earlier Dialectic*. Oxford: Clarendon Press, 1953.

Rosenfeld, Albert. *Second Genesis*. Garden City, N.Y.: Arena Books, 1969.

Ross, W. D. *Aristotle's Physics*. Oxford: Clarendon Press, 1936.

Rostand, E. *Cyrano De Bergerac*. New York: Modern Library, 1951.

Rousselot, Pierre, S.J. *The Intellectualism of St. Thomas*. New York: Sheed and Ward, 1935.

Roszak, Theodore. *Where the Westland Ends*. New York: Doubleday and Company, 1972.

Sagan, C. *Cosmic Connection*. New York: Doubleday and Company, 1973.

________. *Intelligent Life in the Universe*. New York: Dell Publishing Company, 1966.

Saint Exupéry, Antoine de. *The Little Prince*. New York: Harcourt, Brace, Jovanovich, 1943.

________. *Night Flight*. New York: Harcourt, Brace, Jovanovich, 1942.

________. *Wind, Sand and Stars*. New York: Harcourt, Brace, Jovanovich, 1942.

________. *The Wisdom of Sands*. New York: Harcourt, Brace, Jovanovich, 1950.

St. Thomas Aquinas. *De Ente et Essentia*. Critical Text by C. Boyer, S.J. Rome: Gregorian University, 1933.

________. *De Principiis Naturae*. Introduction and Critical Text by John Pauson. Fribourg: Société philosophique, 1950.

________. *De Unitate Intellectus Contra Averroistas*. Critical Text by Leo Keeler, S.I. Rome: Gregorian University, 1946.

________. *In De Anima*. Turin: Marietti, 1935.

________. *In De Caelo*. Rome: Leonine, 1886.

________. *In De Generatione et Corruptione*. Rome: Leonine, 1886.

________. *In Ethicorum ad Nichomachum*. Rome: Marietti, 1934.

________. *In Librum Boethii de Trinitate*. Translation of Questions V and VI by Armand Maurer, C.S.B. *The Division and Methods of the Sciences*. Toronto: The Pontifical Institute of Mediaeval Studies, 1958.

________. *In Librum De Causis*. Rome: Marietti, 1955.

________. *In Metaphysicam Aristotelis*. Turin: Marietti, 1935.

________. *In Peri Hermeneias*. Turin: Marietti, 1955.

________. *In Physicorum Aristotelis*. Rome: Marietti, 1954.

St. Thomas Aquinas. *In Posteriorum Analyticorum*. Turin: Marietti, 1955.

________. *Quaestiones Disputatae De Potentia Dei*. Rome: Marietti, 1931.

________. *Quaestiones Disputatae De Spiritualibus Creaturis*. Rome: Marietti, 1931.

________. *Quaestiones Disputatae De Veritate*. Rome: Marietti, 1931.

________. *Scriptum Super Libros Sententiarum Magistri Petri Lombardi*. Paris: Mandonnet, 1929.

________. *Summa Contgra Gentiles*. Rome: Leonine, 1918–1930.

________. *Summa Theologiae*. Rome: Leonine, 1888–1906.

Sanderson, Ivan. *Invisible Residents*. New York: World Publishing Company, 1970.

Shapley, Harlow. *Of Stars and Man*. New York: Washington Square Press, 1960.

Smith, Vincent. *The General Science of Nature*. Milwaukee: Bruce Publishing Company, 1950.

________. *Philosophical Physics*. New York: Harper and Brothers, 1950.

Stapleton, Olaf. *Last and First Men*. New York: Penguin Books, 1973.

Strughold, Hubertus. *The Green and Red Planet*. Albuquerque, N.M.: University of New Mexico Press, 1953.

Struve, Otto. *Stellar Evolution*. Princeton, N.J.: Princeton University Press, 1950.

Sullivan, Walter. *We Are Not Alone*. New York: New American Library, 1964.

Taylor, A. E. *The Parmenides of Plato*. Oxford: Clarendon Press, 1934.

________. *Plato*. New York: The Dial Press, 1936.

Taylor, Gordon. *Biological Times*. New York: New American Library, 1968.

Toynbee, Arnold. *Study of History*. New York: McGraw-Hill Book Company, Inc., 1972.

Velikosky, Immanuel. *Ages in Chaos*. New York: Doubleday and Company, 1952.

Verne, Jules. *Around the World in 80 Days*. Bridgeport, Conn.: Airmont Publishing Company, Inc., 1964.

Voltaire. *Micromégas: Three Philosophical Voyages*. Germaine Hill, editor. New York: Dell Publishing Company, 1964.

VonBraun, W. *First Men in the Moon*. New York: Holt, Rinehart & Winston, Inc., 1960.

________. *Space Frontier*. New York: Holt, Rinehart & Winston, Inc., 1971.

Von Daniken, Erick. *Chariots of the Gods*. New York: G. P. Putnam's Sons, 1970.

________. *Gods from Outer Space*. New York: G. P. Putnam's Sons, 1968.

________. *The Gold of the Gods*. New York: G. P. Putnam's Sons, 1973.

Webb, Walter P. *The Great Frontier*. Austin: University of Texas Press, 1964.

Whewell, William. *On the Philosophy of Discovery*. New York: Franklin, Burt Publishers, 1971.

Whipple, Fred. *Earth, Moon and Planets*. Cambridge, Mass.: Harvard University Press, 1968.

Wilford, John Noble. *We Reach the Moon*. New York: Bantam Books, 1969.

Wolf, Jack C. and Fitzgerald, Gregor. *Past, Present and Future*. Greenwich, Conn.: Fawcett Publications, Inc., 1973.

Young, Louise R. *Exploring the Universe*. New York: McGraw-Hill Book Company, Inc., 1963.

Zafiropulo, Jean. *Empédocle d'Agrigente*. Paris: Société d'édition "Les Belles Lettres," 1953.

Articles

Allers, R. "The Intellectual Cognition of Particulars." *The Thomist*, Vol. III (1941), pp. 95-163.

Clarke, W. Norris, S.J. "The Limitation of Act by Potency." *The New Scholasticism*, Vol. XXVI, No. 2 (April, 1952), pp. 167–194.

________. "The Meaning of Participation in St. Thomas." *Proceedings of the American Catholic Philosophical Association*, Vol. XXVI (1952), pp. 147–157.

de Koninck, Charles. "La dialectique des limites comme critique de la raison." *Laval théologique et philosophique*, Vol. I, No. 1 (1945), pp. 177–185.

________. "Introduction to the Study of the Soul." *Laval théologique et philosophique*, Vol. III, No. 1 (1947), pp. 9–66.

________. "Les sciences expérimentales sont-elles distinctes de la philosophie de la nature?" *Culture*, Vol. II, No. 4 (Dec., 1941), pp. 465–476.

Dolan, S. Edmund, F.S.C. "Resolution and Composition in Practical Discourse." *Laval théologique et philosophique*, Vol. VI, No. 1 (1950), pp. 9–62.

Drake, F. D. "Project Ozma." *Physics Today*, Vol. XIV (April, 1961), pp. 40–46.

Edelson, Edward. "The Outer Limits of Space." *S/R/World* (October 23, 1973), pp. 66–68.

Foley, Leon A., S.M. "The Persistence of Aristotelian Physical Method." *The New Scholasticism*, Vol. XXVII, No. 2 (April, 1953), pp. 160–175.

Hébert, Thomas, A.A. "Notre connaissance intellectuelle du singulier matériel." *Laval théologique et philosophique*, Vol. V, No. 1 (1949), pp. 33–65.

Holloway, Maurice, S.J. "Abstraction in Cognition." *The Modern Schoolman*, Vol. XXIII, No. 3 (March, 1946), pp. 120–130.

Jansky, K. G. "Electrical Disturbances Apparently of Extraterrestrial Origin." *Proceedings of the Institute of Radio Engineering*, Vol. XXI (Oct., 1933), pp. 1387–1398.

Kane, William H., O.P. "The Extent of Natural Philosophy." *The New Scholasticism*, Vol. XXXI, No. 1 (Jan., 1957), pp. 85–97.

Klubertanz, G., S.J. "St. Thomas and the Knowledge of the Material Singular." *The New Scholasticism*, Vol. XXVI, No. 2 (April, 1952), pp. 35–66.

Lalor, Juvenal, O.F.M. "Notes on the Limit of a Variable." *Laval théologique et philosophique*, Vol. I, No. 1 (1945), pp. 129–150.

Lovell, Bernard. "Search for Voices from Other Worlds." *New York Times Magazine* (December 24, 1961), p. 18.

Marcotte, Normand. "The Knowability of Matter Secundum Se." *Laval théologique et philosophique*, Vol. I, No. 1 (1945), pp. 103–118.

Miller, Marianne. "The Problem of Action in the Commentary of St. Thomas Aquinas on the Physics of Aristotle." *The Modern Schoolman*, Vol. XXIII, No. 4 (May, 1946), pp. 224–226.

Morrison, Philip. "Interstellar Communication." *Bulletin of the Philosophical Society of Washington*, Vol. XVI, No. 1 (1967), pp. 59–81.

Nichol, A. T. "Indivisible Lines." *Classical Quarterly*, Vol. XXX (1936), pp. 120–126.

Nogar, Raymond, O.P. "Towards a Physical Theory." *The New Scholasticism*, Vol. XXV, No. 4 (Oct., 1951), pp. 397–438.

Peters, J. A. J., C.Ss.R. "Matter and Form in Metaphysics." *The New Scholasticism*, Vol. XXXI, No. 4 (Oct., 1957), pp. 447–483.

Pichette, Henri. "Quelques principes fondamentaux de la doctrine du spéculatif et du pratique." *Laval théologique et philosophique*, Vol. I, No. 1 (1945), pp. 52–70.

Smith, Vincent. "Abstraction and the Empiriological Method." *Proceedings of the American Catholic Philosophical Association*, Vol. XXVI (1952), pp. 35–50.

Warren, John. "Nature—A Purposive Agent." *The New Scholasticism*, Vol. XXXI, No. 3 (July, 1957), pp. 364–397.

Zedler, Beatrice. "Averrhoes on the Possible Intellect." *Proceedings of the American Catholic Philosophical Association*, Vol. XXV (1951), pp. 164–178.

Unpublished Material

Beck, Lewis. "Extraterrestrial Intelligent Life." Presidential Address delivered at the meeting of the American Philosophical Association, Eastern Division, December 28, 1971.

Conway, Pierre, O.P. "Thomistic Physics." Unpublished work, College of St. Mary of the Springs, Columbus, Ohio, 1958. (Mimeographed.)

Mullahy, Bernard, C.S.C. "Subalternation and Mathematical Physics." Unpublished doctoral thesis, Department of Philosophy, Laval University, 1947. (Mimeographed.)